Chemistry in the Kitchen Garden

Chemistry in the Kitchen Garden

James R. Hanson
Department of Chemistry, University of Sussex, Brighton, UK

RSCPublishing

ISBN: 978-1-84973-323-6

A catalogue record for this book is available from the British Library

Published by The Royal Society of Chemistry,
Thomas Graham House, Science Park, Milton Road,
Cambridge CB4 0WF, UK

Registered Charity Number 207890

For further information see our web site at www.rsc.org

Foreword

This sequel to Dr Hanson's very well received earlier work, *Chemistry in the Garden*, has been prepared mainly for readers with some chemical background but nevertheless, will also be of considerable interest to anyone wanting to learn more about the natural products that occur in the fruit and vegetables they grow and eat daily.

Apart from the chemistry of these natural products the information provided by the author on the chemistry of the environment in which the fruits and vegetables are grown is very important as it is closely related to the success (or otherwise!) that growers achieve with their crops. Although many gardeners are aware of the importance of soil pH , crop rotation and other factors these are often overlooked when preparation of the soil prior to sowing or planting takes place.

The oft-quoted and much publicised maxim that a healthy diet should include portions of at least five different fruits and vegetables daily has made all of us aware of their very considerable beneficial effects as part of the human diet. We are now in an era where more and more people are enthusiastically growing fruit and vegetables—often organically. These natural products play a critical role as part of our everyday diets and it is important, therefore, to ensure that accurate information about the chemical compounds that occur naturally in the fruit and vegetables we grow is available and their beneficial properties understood.

In the later chapters Prof. Hanson has set out very clearly the chemistry of many vegetable crops, herbs, fruit trees and shrubs and the benefits deriving from them which will also be of considerable value to nutritionists as well as gardeners. But look also for the snippets of folklore that relate to some of the herbs—catmint (catnip), *Nepeta cataria*

Chemistry in the Kitchen Garden
By James R. Hanson
© James R. Hanson 2011
Published by the Royal Society of Chemistry, www.rsc.org

has been proven to be chemically attractive to lions; and "rosemary is for remembrance" may well relate to the chemical make-up of this well-known herb.

Chris Brickell CBE, VMH
Former Director RHS Garden,
Wisley and former Director General, RHS

Preface

Over the past decade there has been a resurgence of interest in growing fruit and vegetables in the garden and on the allotment. Part of the driving force behind this is an increased awareness of the health benefits that can be derived from fruit and vegetables in the diet. The 'five helpings a day' dictum reflects the correlation between a regular consumption of fruit and vegetables and a reduced incidence of, for example, cardiovascular disease and some cancers. Growing your own vegetables provides the opportunity to harvest them at their peak, to minimize the time for post-harvest deterioration prior to consumption and to reduce their 'food miles'. It also provides an opportunity to grow interesting and less common cultivars. The combination of economic advantages and recreational factors add to the pleasure of growing fruit and vegetables.

This book is concerned with the natural products that have been identified in common 'home-grown' fruit and vegetables and which may contribute to their organoleptic and beneficial properties. The book is aimed at readers with a chemical background who wish to know a little more about the natural products that they are eating, their beneficial effects and the roles that these compounds have in nature. It develops in more detail the relevant sections from the earlier RSC book 'Chemistry in the Garden'.

Inspection of any seed catalogue will reveal the many different cultivars of the common fruits and vegetables that are available. Some cultivars have increased resistance to diseases or pests, whilst others have different times of harvest, sizes, colours or other physiological features. Any gardener will also know that the soil and climatic conditions can modify the productivity of fruit and vegetables. Clearly these

Chemistry in the Kitchen Garden
By James R. Hanson
© James R. Hanson 2011
Published by the Royal Society of Chemistry, www.rsc.org

differences in the cultivars and in their cultivation are likely to be reflected in their chemical constituents. The taste and colour of the many varieties of apples provide a good illustration of this. Nevertheless, there are particular compounds that are characteristic of a species and whose presence may be of benefit to man. These form the subject of this book.

Many of the compounds which we value for their contribution to the organoleptic properties of fruit or vegetables have a specific ecological role to play in the development and protection of the plant. Some may be produced as insect deterrents or as inhibitors of fungal infection and they may be produced as a consequence of particular stimuli. The variations in the natural product content of fruit and vegetables as they reach maturity and ripen have a considerable impact on their palatability. In the context of the chemical composition of fruit and vegetables, it is important to consider the role that the relevant part of the plant may be playing in its overall life-cycle. The edible parts of some vegetables may be their tuberous storage organs, whilst others are part of the synthetic apparatus of the leaves and yet others are the seeds or the parts that contain the seeds.

Unlike other herbivorous mammals, man has learnt to prepare fruit and vegetables for consumption by, for example, boiling them in water. This ability to convert unpalatable and indigestable plant material into something edible gave early man a competitive advantage over other herbivores in seeking out foodstuffs. There is an interesting observation in this connection. In an examination of over two thousand wild plants, approximately 11% were found to be cyanogenic, yet of the top 24 food crops, 16 (*i.e.* two-thirds) were cyanogenic. Cyanogenesis protects the plant against substantial herbivore consumption, whilst the preparation of food by man can serve to detoxify it thus favouring the domestication of particular plants. Furthermore, man has transported and domesticated plants from around the world thus removing them from their natural predators. Apart from the gains in this process of domestication and the development of cultivars, there have been losses not only of some protective characteristics but also of some traits which we now know to be beneficial. The loss of beneficial anthocyanins from the original purple carrot to create the orange-coloured carrot is a case in point. The preservation of seed banks of the 'wild' plant and its relatives, together with an understanding of their constituents, is an important project.

The post-harvest treatment of fruit and vegetables can significantly alter their constituents and potential effects. The post-harvest persistance of a number of enzymes which modify plant constituents such as alliinase and the onion lachrymator factor synthase in onions, the myrosinase in cabbage and the phenolases or polyphenol oxidases in potatoes and apples, are examples. In some food processing, a rapid blanching will destroy these. Many of the beneficial polyphenolic compounds, vitamins and glycosides have an appreciable solubility in hot water and may be lost during cooking. The use of vegetable stock for making soup or gravy is a way of retaining some beneficial compounds in the food chain. The use of olive oil with many herbs is helpful because terpenoid and aromatic compounds are often lipid soluble and are eluted from the leaves by the oil and retained in the food. Storage is another area where post-harvest changes including cell wall degradation, the loss of water, microbial decay and sprouting may occur. The chemistry and chemical control of these features can prolong the period over which home-grown vegetables and fruit may be used.

The book begins with an outline of the major groups of compound that are found in fruit and vegetables. This is followed by a description of aspects of environmental chemistry that contribute to the successful cultivation of these crops. Subsequent chapters deal with individual plants which are grouped in terms of the part of the plant – roots, bulbs and stems, leaves and seeds – that are used for food. The final chapters deal with fruit and herbs.

Developments in the understanding of the ecological and beneficial chemistry of fruit and vegetables have made the exploration of their chemical diversity a fascinating and expanding area of natural product chemistry. It is hoped that the reader will obtain a 'taste' for this chemistry from this book.

I wish to thank Dr Merlin Fox and other members of the staff of the Books Department of the Royal Society of Chemistry for their help in the production of this book.

Contents

Chemistry in the Kitchen Garden
By James R. Hanson
© James R. Hanson 2011
Published by the Royal Society of Chemistry, www.rsc.org

CHAPTER 1

Natural Products in Fruit and Vegetables

1.1 INTRODUCTION

The constituents of edible plants have been the subject of investigation throughout the development of chemistry. Many of the major carbohydrates, fatty acids, amino acids, mineral and vitamin constituents of fruits and vegetables together with their pigments and flavours were isolated and identified during the nineteenth and the first half of the twentieth centuries. In the second half of the twentieth century the advent of instrumental methods of separation and analysis, such as gas chromatography linked to mass spectrometry, high pressure liquid chromatography, ultra-violet, infra-red and nuclear magnetic resonance spectroscopy, permitted far more detailed investigations. Furthermore, the realization that particular components of fruit and vegetables conferred specific health benefits provided the stimulus for more extensive investigations to identify the bioactive natural products.

The organic compounds that occur in plants fall into three main groups. Firstly, there are the high molecular weight polymeric materials, such as cellulose, starch and lignin, which, together with various proteins and nucleic acids, form the structural, storage, enzymatic and genetic components of the cell. Secondly, there are those compounds of lower molecular weight that occur in the majority of plant cells and which play a central role in the metabolism and reproduction of the cell. These are sometimes known as the 'primary metabolites'. They include the common sugars, some carboxylic acids and the amino acids that are the

Chemistry in the Kitchen Garden
By James R. Hanson
© James R. Hanson 2011
Published by the Royal Society of Chemistry, www.rsc.org

constituents of peptides and proteins. There are also heterocyclic compounds that are co-enzymes and others which form part of the nucleic acids. Related to these are the plant hormones and signalling compounds which regulate the overall growth and development of the plant. The third group of naturally-occurring compounds also includes relatively low molecular weight compounds, are those which are characteristic of a limited range of species. These compounds may have insect attractant or deterrant roles, or they may provide a defense against microbial attack. They serve to establish an ecological niche for the plant. These natural products can behave as 'semiochemicals' which convey a chemical message between species. In plants, many of these compounds are defensive 'allomones' being produced to benefit the source but to the detriment of the receiver, typically an insect herbivore. These natural products are sometimes known as 'secondary metabolites' and include many of the compounds that are responsible for the particular health benefits of specific fruits or vegetables as well as their colour and flavour. In this context the organoleptic and beneficial properties of fruits and vegetables are often the summation of contributions from many compounds. The naturally occurring compounds in foodstuffs that are beneficial to man are sometimes called 'nutraceuticals'. Although these primary and secondary metabolites make significant contributions to plant biochemistry and ecology, they are often present in very small concentrations in the plant, typically milligrams per kilogram of fresh weight.

Whereas many of the primary metabolites exert their biological effect within the cell in which they are produced, the secondary metabolites often exert their biological effects on other cells or species. However, this division between primary and secondary metabolites, whilst useful, is not rigid. Acids derived from primary metabolism form esters with secondary metabolites, amino acids that are constituents of proteins are also the progenitors of the alkaloids and glucosinolates that are secondary metabolites, whilst the common sugars are also found as components of the glycosides of secondary metabolites. The distinction is also blurred in the context of the vitamins and hormones. The former, such as vitamin C, are essential dietary factors for man but are commonly formed by plants. Different plants may produce different but structurally related hormones and use them for similar purposes.

The structures of the secondary metabolites not only vary between species but may also show infra-species variation. Cultivars may be bred for different purposes such as disease resistance, colour, size or taste, all factors which may be determined by their secondary metabolite content. Furthermore, the chemistry of a plant changes as it develops and approaches maturity, as exemplified by the colour and taste of many fruit as they ripen. Some secondary metabolites are also produced as a consequence of external pressures, such as attack by a fungus or herbivore. These variations lead to what may seem at first sight to be a bewildering array of natural products. However, the structural diversity of the natural products that are found in fruit and vegetables can be rationalized in terms of their biosynthesis. A map illustrating these relationships and following the pathway of carbon from carbon dioxide is shown in Figure 1.1. It is helpful to use this scheme as a framework against which to describe the various groups of natural product.

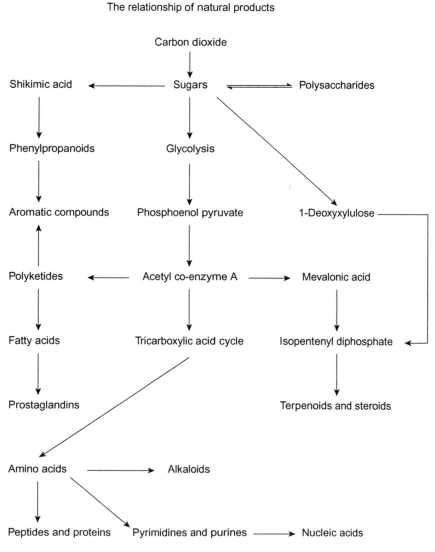

The relationship of natural products

Figure 1.1

1.2   THE BIOSYNTHETIC RELATIONSHIP OF NATURAL PRODUCTS

Carbon dioxide from the atmosphere is incorporated by photosynthesis into sugars from which the plant derives some of its structural and storage materials, such as cellulose and starch. Compounds of lower molecular weight may be derived from the sugars. Thus the breakdown of the C_6 sugar, fructose 1,6-diphosphate **1.1**, by glycolysis affords two C_3 fragments, dihydroxyacetone monophosphate **1.2** and glyceraldehyde monophosphate **1.3**, through the biochemical equivalent of a retro-aldol reaction. The C_3 units provide the source of pyruvic acid **1.4** and thence the acetate units from which many natural products are formed. Acetyl co-enzyme A **1.5** also enters the tricarboxylic acid cycle from which the common plant acids, citric and malic acids, are formed.

   Although there are a very large number of natural products, their carbon skeleta are assembled from relatively few elementary building blocks such as acetyl co-enzyme A **1.5**, isopentenyl diphosphate **1.6** and shikimic acid **1.7**. Pathways involving transamination lead to the formation of amino acids from keto acids and the inclusion of nitrogen into other compounds. The various natural products are then formed by biosynthetic pathways in which these building blocks are linked together, cyclized and oxidized. In many plants it is possible to identify the

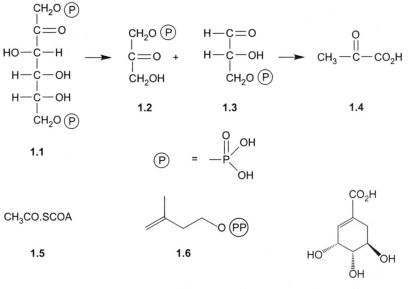

groups of natural products which they contain, and place these in sequences which reflect their various biosynthetic pathways.

The biosynthetic rationalization of structures not only inter-relates natural products but also reflects the botanical relationships between plants. The occurrence of a particular biosynthetic pathway in a plant indicates the presence of particular enzymes which catalyze steps in the pathway and these in turn reflect the genetic make-up of the plant. Not surprisingly, therefore, botanically related plants produce similar natural products. As we will see in subsequent chapters, the plant families to which various fruits and vegetables belong are often characterized by the presence of particular types of natural product.

1.3 SUGARS

The sugars provide the basic building blocks for the polysaccharide structural materials of fruits and vegetables. The two most widely occurring C_6 monosaccharides are the aldose, glucose **1.8**, and the ketose, fructose **1.10**. The C_5 sugars, ribose and deoxyribose, are found in the nucleic acids. Other sugars may be found in glycosides. The majority of naturally-occurring sugars are all related to one enantiomeric form (the D-form) of glyceraldehyde.

The sugars exist in their cyclic hemi-acetal form. In the case of glucose **1.8** this is normally the six-membered pyranose ring, in which all of the hydroxyl groups have the equatorial conformation. Fructose **1.10** is also found in a five-membered furanose form. Another important property of the sugars is their ability to form ether linkages between the hemi-acetal hydroxyl group of one monosaccharide and a hydroxyl group of a second sugar to give a di- and eventually a polysaccharide. Cellobiose **1.11** is a disaccharide derived from a β-1,4 bond between two glucose molecules and which forms the repeating unit of cellulose. Sucrose **1.12**, on the other hand, has an α-1,2 link between a glucose and the furanose form of fructose. The ether link between the two hemi-acetal carbons of glucose and fructose masks a potentially reactive centre in each of the monosaccharides restricting some of the reactivity of sucrose and contributing to its role as a useful biological energy storage compound. Maltose has an α-1,4 linkage between two glucose units. In the context of foodstuffs, the importance of the nature of these linkages lies in the variation with which the different glycosidic linkages undergo enzymatic cleavage in man, thus affecting our ability to digest foods.

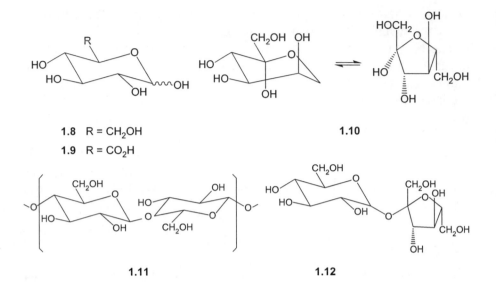

1.8 R = CH₂OH

1.9 R = CO₂H   **1.10**

1.11   **1.12**

1.4   STRUCTURAL AND STORAGE POLYSACCHARIDES

Cellulose is one of the most abundant bioorganic polymers. It is a β-1,4′-polyacetal of the disaccharide cellobiose **1.11**. The number of sugar units in the polymer may be as high as 10–15 000. Due to the many opportunities for both intra- and inter-molecular hydrogen-bonding between the sugar units and the summation of the energy that this represents, a network is created that hinders free rotation and imposes rigidity on the structure. Since the bonding between the sugars involves equatorial rather than axial hydroxyl groups, the polysaccharide adopts an extended linear conformation to give fibrous structures. This provides the basis for the function of cellulose as a structural component of leaves and other parts of the plant.

Although the polymeric structure of cellulose precludes water solubility, the extensive hydrogen-bonding network and the presence of additional sites for hydrogen-bonding mean cellulose can absorb a significant amount of water. This interaction with water confers both advantages and disadvantages to the plant. Since the cellulose in leaves can absorb water, the surface of the leaf is often covered with a hydrophobic hydrocarbon wax to reduce waterlogging when it rains. This wax coating may also prevent excessive water loss from the plant under arid conditions. The development of cultivars with this wax protection to the leaf may be important if we enter a period of long dry summers through global warming.

This wax also behaves as a lubricant and in the autumn it makes freshly fallen wet leaves slippery. When the leaf has died, biodegradation of the wax eventually exposes the cellulose of the leaf which can then become waterlogged and subject to microbial attack. The wax coating on fruit is an important protection against dehydration and microbial spoilage. Apples with a good wax coating keep longer. There was an old method of storing apples which used an oiled paper wrapping to reduce spoilage.

Whereas cellulose is a structural polysaccharide, starch is a storage polysaccharide. Thus starch is the major carbohydrate reserve in potatoes, whilst leafy plants, such as lettuce, contain relatively little starch and more cellulose. Starch differs from cellulose by containing α-1,4' linked D-glucose units. These form a polymer known as α-amylose. The other major component of starch is amylopectin which contains the α-1,4' linked D-glucose backbone together with some chain branching. Typically starch contains about 20–25% amylose and 75–80% amylopectin. Whereas cellulose has a fibrous structure, the α-amylose of starch with an α-1,4' (axial-equatorial) linkage is wound into a more compact structure. The rigidity of this hydrogen-bonded structure in starch granules is destroyed as water is allowed to penetrate during cooking, for example, of potatoes so softening the vegetable.

Different enzyme systems are involved in the cleavage of cellulose and starch leading to differences in the ease with which they are digested. The human intestine can degrade starch but cellulose is only degraded by the bacteria in the gut and thus much of the cellulose acts as a dietary fibre.

Plants also produce a number of polysaccharides derived from other sugars, such as fructose, galactose, arabinose and xylose. These are known collectively as the 'hemicelluloses'. One particular example is the fructan, inulin, which is produced by chicory and Jerusalem artichokes. Inulin is a polymer of about thirty fructose units linked in a β-2,1 manner. Its bacterial fermentation in the gut produces large quantities of short-chain fatty acids and gases, such as carbon dioxide and methane, which cause discomfort when eating foods with high inulin content.

When a tree or a fruit is damaged, it may form a gummy exudate to protect the site of injury. These plant gums are polysaccharides which possess a branched chain structure often containing a glucuronic acid **1.9** unit. A typical plant gum might contain a core of β-1,3 D-galactose units to which side chains of different sugars such as L-arabinofuranose, L-rhamnopyranose and D-glucuronic acid **1.9** are attached. This branched chain structure reduces the tendency of the gum to crystallize and, because there are different sugars requiring different enzymes to metabolize them, it also reduces the susceptibility of the gum to microbial breakdown. The mucilage around germinating seeds is also formed of polysaccharides. The hydrogen-bonding properties of the sugar units helps to retain water and protect the seeds from dessication, and provide an initial channel for the uptake of water and nutrients. The pectins which are formed in the primary cell walls of some fruits, such as apples, contain D-galactouronic acid units which are partially methylated. These provide the basis for the extensive gelling properties of the pectins which occur in the presence of plant acids and sugar in jam making.

1.5   LIGNIN

As a plant ages, lignin begins to permeate the polysaccharide membrane. Lignin comprises aromatic C_6-C_3 units and it is structurally quite different from the polysaccharides. The biosynthesis of the C_6-C_3 building block through shikimic acid, is described later. Lignin is formed from building blocks of 4-hydroxyphenylpropenol **1.13**, coniferyl alcohol **1.14** and sinapyl alcohol **1.15** in which the aromatic rings confer rigidity. The polymerization process is a free-radical, phenol-coupling reaction. These oxidative processes are mediated by iron-containing heme enzymes and lead to structures containing units such as **1.16**.

1.13 $R^1 = R^2 = H$

1.14 $R^1 = OMe, R^2 = H$

1.15 $R^1 = R^2 = OMe$

The radicals derived from compounds such as coniferyl alcohol have several sites for phenol-coupling reactions allowing for cross-coupling between chains to occur, imparting strength, for example, to wood. These enzymatic phenol-coupling reactions also lead to low molecular weight compounds known as 'lignans', such as pinoresinol **1.17**. Some of these compounds have attracted interest because of their biological activity as antitumour agents. Since these compounds are electron-rich phenols, they are also powerful antioxidants.

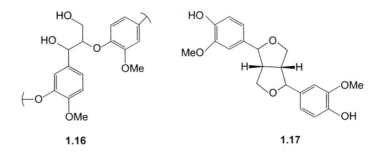

1.16   **1.17**

1.6 LOW MOLECULAR WEIGHT NATURAL PRODUCTS

The central biosynthetic pathway in Figure 1.1 shows that there are various major building blocks for natural products which are derived by the metabolism of sugars. The majority of the natural products which are known as 'secondary metabolites', belong to one of several large families of compounds whose structural characteristics reflect the basic building blocks from which they are assembled. These classes are:

- Fatty acids and polyketides
- Terpenoids and steroids
- Phenylpropanoids and flavonoids
- Alkaloids
- Specialized amino acids and their derivatives
- Specialized sugars.

Whilst the division of these natural products into families depending on their building blocks is convenient, it is not absolute. Thus there are a number of natural products in which part of the structure may be assembled by the phenylpropanoid (C_6-C_3) pathway and other parts are derived by the polyketide or terpenoid pathway.

1.7 FATTY ACIDS AND POLYKETIDES

Long-chain fatty acids occur in varying extents throughout the plant kingdom, particularly in the wax coating of leaves, in cell membranes and in seed oils. Vegetables and fruits are an important dietary source of unsaturated fatty acids, such as oleic acid **1.18**, linoleic acid **1.19** and the triene, linolenic acid. The acids occur in lipid fractions often attached to glycerol as triacylglycerol esters. They are also constituents of glycolipids and phospholipids. The related long-chain alcohols are also found. These fatty acids are biosynthesized from acetyl co-enzyme A *via* malonate in a sequence of carbanion condensations and reductions. This involves the formation of a saturated fatty acid which is then followed by *cis* dehydrogenations to generate the (Z)-alkene of the unsaturated fatty acid. A characteristic feature of the polyunsaturated fatty acids is that the double bonds are not conjugated but are separated by a methylene. Importantly the double

1.18	**1.19**

bonds, as in oleic acid **1.17**, have the *cis* geometry. Although the *trans*-isomers, for example elaidic acid, do occur naturally, they are more often formed by the chemical hydrogenation and isomerization of polyunsaturated fatty acids. The *trans* acids and their derivatives crystallize more easily and are less rapidly metabolized than the *cis* acids leading to unwanted deposits of fat in the body.

Since the fatty acids are formed from the two carbon unit of acetyl co-enzyme A, they possess an even number of carbon atoms. Whilst a significant number are C_{18} acids, others have a C_{16}, a C_{20} or even a C_{22} chain. Thus the C_{22} erucic acid [(Z)-docos-13-enoic acid] is found in members of the Brassicaceae. In man, the C_{20} polyunsaturated acid, arachidonic acid [(all *Z*)-eicosatetra-5,8,11,14-enoic acid], is the pre-cursor of the prostaglandins and related hormones. It is formed from linoleic acid by the introduction of two more double bonds and exten-sion of the carbon chain. The plant hormone, jasmonic acid, has a structure **1.20** which is reminiscent of the prostaglandins. Further enzymatic dehydrogenation reactions lead to acetylenic compounds which are found in a number of vegetables, such as carrots and parsnips. They are often formed as a consequence of microbial attack.

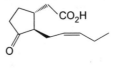

1.20

Oxidative cleavage of the unsaturated chains by the oxylipin pathway leads to the aldehydes, hex-3-enal and hex-2-enal, which make important contributions to the aroma of many vegetables. Under normal conditions, regulation of this oxidative degradation is maintained by the presence of antioxidants such as the carotenoids and the tocopherols (vitamin E) in the lipid fraction. However, disruption of the cells on the surface of a plant leads to the oxidative degradation of the unsaturated fatty acids. Decarboxylation of some of the longer chain acids also affords hydro-carbon waxes which can form part of the hydrophobic coating of the leaves. These waxes also include long-chain alcohols derived by reduction of the acids, and these in turn form esters with the fatty acids.

If the reduction of the polyketide carbonyl groups arising from the initial condensation reactions in the biosynthesis of the fatty acid chain does not occur, these carbonyl groups may participate in a number of cyclization reactions leading to aromatic and heterocyclic compounds. These polyketide pathways have been thoroughly studied in the context of microbial metabolites. However, they also often occur in plants in conjuction with other pathways, such as the phenylpropanoid pathway. In an important example, the phenylpropanoid cinnamate unit acts as a starter for the addition of three acetate units to form the chalcones and flavonoids which are found in many fruits and vegetables. These are discussed later (section 1.9).

A number of other acids which arise from the central primary metabolism involving acetyl co-enzyme A in their biosynthesis make an important contribution to the taste of fruit and vegetables. These include citric acid **1.21**, malic acid **1.22** and tartaric acid **1.23**. The citric acid (Krebs tricarboxylic acid) cycle is a major metabolic pathway in glucose catabolism, releasing carbon dioxide and energy and generating building blocks for amino acid and porphyrin biosynthesis.

$$CH_2CO_2H \quad\quad CH_2CO_2H \quad HO—CH\,CO_2H$$
$$HO—C—CO_2H \quad H—C—OH \quad HO—CH\,CO_2H$$
$$CH_2CO_2H \quad\quad CO_2H$$

$$\textbf{1.21} \quad\quad\quad \textbf{1.22} \quad\quad\quad \textbf{1.23}$$

1.8 TERPENES AND STEROIDS

The carbon skeleton of the terpenes and steroids is assembled from a C_5 building block, isopentenyl diphosphate **1.25**, which provides the characteristic branched chain isoprene unit that is the hallmark of the terpenes and steroids. Isopentenyl diphosphate is formed in plants by two separate biosynthetic pathways. One involves three acetate units and proceeds *via* the C_6 compound, mevalonic acid **1.24**. The mevalonate pathway, which is also found in mammals and fungi, was first established over fifty years ago. The second, more recently discovered pathway proceeds directly from sugars *via* 1-deoxyxyulose monophosphate **1.26**. This pathway was first identified in bacteria but has since been found in many plants. Whereas the 1-deoxyxyulose pathway is commonly found in the plastids, the mevalonate route is more typical of the cytosol.

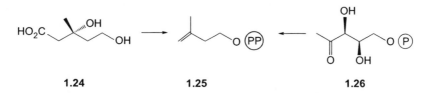

$$\textbf{1.24} \quad\quad\quad\quad \textbf{1.25} \quad\quad\quad\quad \textbf{1.26}$$

The terpenes can be divided into groups depending on the number of constituent isoprene units. Thus, the C_{10} monoterpenes contain two isoprene units linked in a head-to-tail manner. The C_{15} sesquiterpenes are formed from three units, the C_{20} diterpenoids four units, the C_{30} triterpenoids six units and the C_{40} carotenoids eight units. Rubber is a polyisoprenoid compound. The C_{30} triterpenoids are the precursors of the plant sterols and the steroid hormones. In addition, an isoprene unit is often found attached to a natural product that is formed by another pathway.

The monoterpenes include many compounds which contribute to the flavour and aroma of herbs, fruit and vegetables. In the plant they play a major ecological role in mediating interactions between plants and insects, and in allelopathic interactions between plants. They are found in all parts of the plant, often as complex mixtures. In the leaves they may be stored in glandular trichomes or hairs and released when the leaves are damaged. We often crush a leaf, such as mint, between our fingers to release the odour. Although one monoterpene may predominate and provide the major impact, the overall aroma of a plant is usually the summation of many constituents which may include volatile compounds derived from other biosynthetic pathways. Gas chromatography has played an important role in the separation of these mixtures. The detailed composition of these mixtures may vary from one cultivar to another, with the maturity of the plant and with the climate and even with the location, a feature commonly observed with the mints. In the Middle Ages, the walled gardens of a monastery often provided a micro-climate which favoured the growth of medicinal herbs, many of which produce monoterpenes. Culinary herbs related to these often grow well in pots in sheltered sunny positions on a patio or even on a window sill.

The monoterpenes can be divided into acyclic (*e.g.* geraniol **1.27**), monocyclic (*e.g.* menthol **1.28**) and bicyclic (*e.g.* α-pinene **1.29**) compounds. Within each group there are hydrocarbons, alcohols, aldehydes and ketones. Some of the alcohols may be stored in the plant as their glycosides. There are many closely related monoterpenes which may be structural, stereochemical or enantiomeric isomers of each other. The perception of odour is a response to a combination of compounds binding to different receptors. Since the proteins of these receptors have specific conformations and utilize chiral amino acids, it is not suprising that different isomers have different odours. Indeed odour is enantiospecific. For example, *R*-(−)-carvone **1.30** which occurs in spearmint (*Mentha spicata*) has a peppermint odour, whilst its enantiomer, *S*-(+)-carvone **1.31**, found in caraway and dill has an odour characteristic of these herbs. The menthol series often occurs in various stereoisomeric forms.

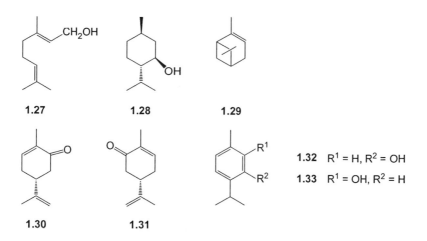

1.27   1.28   1.29

1.30   1.31

1.32 R^1 = H, R^2 = OH

1.33 R^1 = OH, R^2 = H

There are other groups of monoterpenes which have a cyclopentanoid skeleton. These are less common in garden plants although they are found in olives, antirrhinums and in weeds such as plantain, as well as in some insects.

Apart from contributing to the aroma of plants, many monoterpenes behave as semiochemicals affecting interactions between plants, and between plants and herbivores or plant pathogenic micro-organisms. Indeed the antimicrobial action of the phenolic monoterpenes, thymol **1.32** and carvacrol **1.33**, extends to the control of human pathogenic organisms and those that cause food spoilage. Both thymol and carvacrol are found in proprietary medicines (glycothymolein throat pastilles and carvacrol decongestant vaporizers).

The sesquiterpenes contain three isoprene units and are formed by various modes of cyclization of the C_{15} farnesyl diphosphate **1.34**. The more volatile sesquiterpenes, such as humulene, caryophyllene **1.35**, eudesmol **1.36** and β-cadinene **1.37**, contribute to the aroma of fruit and vegetables. They can act as insect attractants. Many insects respond to the farnesenes which are dehydration products of farnesol. The insect juvenile hormones are homologues of epoxyfarnesol derivatives. The more highly oxygenated sesquiterpenoids include a large number of lactones, some of which have been identified as the bitter principles found in crops such as lettuces. They are particularly widespread in plants that are members of the Asteraceae (Compositae) family. The toxic trichothecenes, such as deoxynivalenol (DON), are metabolites of food spoilage fungi belonging to the *Fusaria*. The plant hormone, abscisic acid **1.38**, which brings about leaf fall in the autumn and in times of water stress is a sesquiterpenoid.

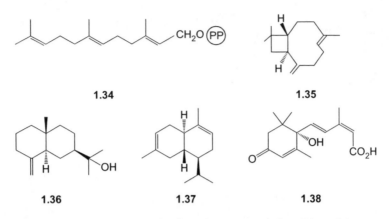

1.34   **1.35**

1.36   **1.37**   **1.38**

The C_{20} diterpenoids contain four isoprene units. Phytol is an acyclic diterpenoid unit which is attached as an ester to the porphyrin skeleton of chlorophyll. However, the majority of the diterpenoids are bi-, tri- or tetra-cyclic substances. Bicyclic diterpenoids of both the labdane and the rearranged clerodane series are quite common constituents of the Lamiaceae (Labiatae), such as sage. Some of these compounds have powerful insect antifeedant activity. Sage and rosemary also contain tricyclic diterpenoid phenols, such as carnosol **1.39**, which are powerful antioxidants. These phenols are also antimicrobial agents and contribute to the use of these herbs in preserving meats. The resin acids which occur in many pines are tricyclic diterpenoids. The gibberellin plant hormones, *e.g.* gibberellic acid **1.40**, which mediate many aspects of plant growth and development, are tetracyclic diterpenoids. As with the mono- and sesquiterpenoids, the diterpenoids occur in both enantiomeric series.

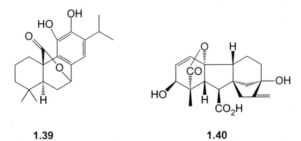

1.39   **1.40**

There are relatively few C_{25} sesterterpenoids that are found in higher plants, but the C_{30} triterpenoids are widespread in fruit and vegetables, as well as in other plants. Although these contain six isoprene units, they are formed by the tail-to-tail linkage of two C_{15} farnesyl diphosphate units to give the parent hydrocarbon, squalene. In plants, the cyclization of squalene monoepoxide affords tetra- and penta-cyclic triterpenes. Whilst cycloartenol **1.41** is formed by the cyclization of squalene monoepoxide in many plants, lanosterol is the precursor of the sterols in mammals and fungi. Its conversion into the fungal sterol, ergosterol, is the target of the azole and morpholine fungicides, such as triadimeton® and tridemorph®. The tetracyclic triterpenes form the precursors of the plant sterols, such as stigmasterol and sitosterol. The plant sterols have a beneficial effect in the diet by reducing the amount of the animal sterol, cholesterol, in the circulatory system. The cucurbitanes are rearrangement products of the tetracyclic triterpenes and are found in cucumbers and pumpkins. They contribute to the bitter taste of some cucumbers. Pentacyclic triterpenes, such as oleanolic acid **1.42**, are found in the wax coatings of some leaves and fruit, as well as in the bark of trees. They and some of the steroids can occur as surface-active glycosides (saponins) in plants and will generate a foam in aqueous solution. The name 'saponin' is derived from the Latin '*sapo*' for soap. The foam found on some plants enveloping invasive insects is created by saponins produced by the plant. Some of the pentacyclic triterpenes are attracting interest for their anti-inflammatory activity, whilst some tetra- and penta-cyclic triterpenes have tumour inhibitory activity. Unlike the lower terpenoids, the naturally occurring triterpenoids and steroids belong to one enantiomeric series.

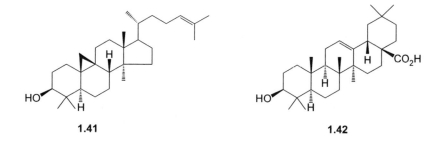

1.41   **1.42**

The C_{40} carotenoids contain eight isoprene units and are formed by the tail-to-tail linkage of two C_{20} geranylgeranyl diphosphate units. These compounds are important pigments, for example, in tomatoes and carrots. They have a useful light-absorbing and antioxidant role. Vitamin A is formed by the cleavage of β-carotene **1.43**. Some of their other degradation products, such as β-ionone, make a useful contribution to the aroma of plants.

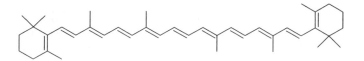

1.43

1.9   PHENYLPROPANOID COMPOUNDS

The phenylpropanoid family of natural products is that which contains compounds with a C_6-C_3 unit. There are three large groups of compounds found in fruit and vegetables which contain this unit. The first are aromatic acids, such as cinnamic acid, and oxygen heterocycles, such as the coumarins. The second are the flavonoids which include many plant pigments, and the third are amino acids, such as phenylalanine, tyrosine and certain alkaloids and their relatives which are derived from them. Many of these compounds are phenols and are easily oxidized. A number of the beneficial properties of the phenylpropanoids in fruit and vegetables have been associated with their interactions with reactive oxygen species.

The biosynthetic pathway leading to the C_6-C_3 unit is known as the shikimic acid pathway. The pathway falls into two sections. The first involves the formation of the cyclic shikimic acid **1.46** from phosphoenol pyruvate **1.44** and the C_4 sugar, erythrose 4-phosphate **1.45**, *via* the phosphate of a C_7 acid, heptulosonic acid. The second stage involves the addition of a further phosphoenolpyruvate to shikimic acid to give chorismic acid **1.47**. This undergoes a Claisen rearrangement to prephenic acid **1.48**. Decarboxylation then generates the aromatic ring of phenylpyruvic acid **1.49** from which the amino acid, phenylalanine **1.50**, and the unsaturated acid, cinnamic acid **1.51**, are derived.

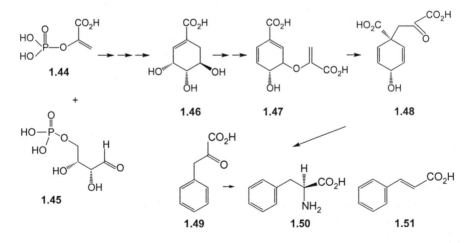

This pathway is not found in man, and thus phenylalanine, tyrosine and tryptophan which are derived by this pathway are essential dietary amino acids. A key enzyme in the pathway in plants is 5-enolpyruvylshikimate-3-phosphate synthase (EPSP synthase) which catalyses a step between shikimic acid and chorismic acid. This enzyme is the target for the herbicide, glyphosate (Roundup®) **1.52**.

$$HO \diagdown P(=O) \diagup HO \quad CH_2\,NH\,CH_2\,CO_2H$$

1.52

Metabolites from the phenylpropanoid pathway can combine with those from other pathways. Hydroxycinnamic acids, such as *p*-coumaric acid, caffeic acid, ferulic acid and sinapic acid, often occur as esters of quinic acid and are collectively known as the chlorogenic acids, *e.g.* chlorogenic acid **1.53**. Rosmarinic acid **1.54** which is found in a number of plants, particularly culinary herbs of the Lamiaceae, is an ester of caffeic acid with 2',3,4-trihydroxyphenylpropionic acid. These compounds have valuable antioxidant properties. The presence of adjacent oxygen functions and the aromatic ring leads to their ability to deactivate some reactive oxygen species, scavenge free radicals and complex with metal ions involved in oxidative processes.

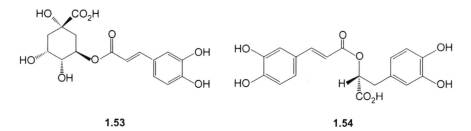

1.53   **1.54**

Coumarins are also formed by the C_6-C_3 pathway. These phenols can be prenylated on the aromatic ring, and on cyclization these isoprenoid fragments generate furanocoumarins, such as psoralen **1.55**. These compounds are characteristic of the Apiaceae (Umbelliferae) and are the phototoxic metabolites of, for example, parsnips. Gallic acid (3,4,5-trihydroxybenzoic acid) is produced by the shikimate pathway. It can be reversibly linked to glucose in the ellagitannins. Gallic acid is also found as the dimer, ellagic acid, an antioxidant which is found in numerous fruits and vegetables.

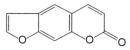

1.55

The flavonoids are C_6-C_3-C_6 compounds containing two aromatic rings connected by a C_3 bridge. They are biosynthesized through a combination of the C_6-C_3 and polyketide pathways. The aromatic ring C posesses an oxygenation pattern characteristic of the phenylpropanoid pathway, whilst the other aromatic ring has oxygen functions on alternate carbon atoms typical of a polyketide pathway. The various groups of flavonoid differ from each other in the oxidation level of the central oxygen heterocycle. Important flavonoids are the flavones **1.56**, the flavan-3-ols **1.57**, the flavonols **1.58** and the anthocyanidins **1.59**. The isoflavones **1.60** are rearrangment products which are characteristic of the legumes. Reduction products such as the flavanones, dihydro-flavonols and flavan-3,4-diols are also found in nature. The chalcones lack the heterocyclic ring. Most of the flavonoids are found as glycosides with a range of different sugars attached to the hydroxyl groups. Some of the sugars may also be acylated. Although the number of parent aglycones is relatively small, the overall number of flavonoid glycosides is very high. The glycosides of the anthocyanidins are known as the 'anthocyanins'.

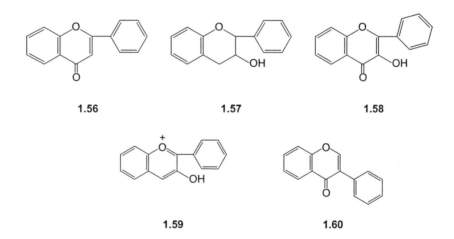

1.56   1.57   1.58

1.59   1.60

The flavonoids are located in the epidermal layer of leaves, in flowers and in the skin of fruits where their light-absorbing properties and antioxidant activity can protect vunerable parts of the plant from excess ultraviolet radiation and oxidative stress. An important feature of their structures is the stabilization provided by the conjugated system to free radicals formed by the oxidation of phenols. Although the light absorption of the anthocyanins is pH dependent, the contribution of these pigments to the colour of the plants arises from co-pigmentation between stacked molecules of a flavone and an anthocyanin, in which the overall complex is held together by metal ions such as iron, magnesium and calcium.

Dietary flavonoids are associated with a valuable antioxidant activity. They also have a beneficial effect in reducing chronic inflammation which can predispose the body to more serious human diseases including cancer. Chronic inflammation has been linked to cardiovascular disease, neurological and metabolic disorders, and bone and muscular diseases, many of which are associated with ageing processes. For example, the electron-rich flavonoid, luteolin **1.61**, has a very significant radical scavenging and antioxidative ability. This can be associated not only with the delocalization of phenoxide radicals **1.62** but also with the stabilization of different oxygen radicals by hydrogen bonding. These compounds suppress the formation of pro-inflammatory cytokines.

Isoflavonoids, such as genistein **1.63**, and the coumestan, coumestrol, have a formal similarity to the steroid hormone, estradiol. They bind to the estrogen receptor and are weakly estrogenic. Consequently they are known as 'phytoestrogens'. Apart from this biological activity, they also show antiviral activity.

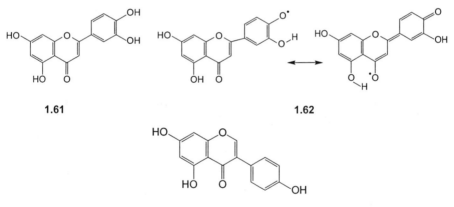

1.61   **1.62**

1.63

Polymeric forms of the flavonoids known as 'proanthocyanidins' also occur. In these, there is a bond between C-8 of one molecule and C-4 of a second molecule. The role of the C_6-C_3 pathway in the formation of lignin has been mentioned earlier.

1.10 AMINO ACIDS

The essential amino acids that form the constituents of proteins are found in plants. Whilst the twenty two proteogenic amino acids are found throughout the plant kingdom, there are a further large number of non-proteogenic amino acids found in particular plants and which have specialist, often allelochemical roles. Many amino acids are formed by transamination of the corresponding keto acids which play a central role in plant metabolism. However, in some cases, such as the biosynthesis of L-leucine and L-valine, the pathway can be quite lengthy. Amino acids, such as glycine, provide parts of the nucleic acid bases and, *via* δ-aminolaevulinic acid **1.64**, the porphyrin ring system of chlorophyll and the heme pigments. Some amino acids and their derivatives such as asparagine and glutamine **1.65** were first isolated from vegetables, the former from asparagus and the latter from sugar beet. Glutamine plays an important role in the assimilation and storage of ammonia in plants. Tryptophan is a precursor of the auxin plant hormone, indolylacetic acid. In addition, there are some amino acid derivatives such as (*S*)-2-propenylcysteine *S*-oxide (alliin) which have a particular role, in this case as precursors of sulfenic acids in *Alliums*.

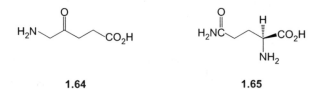

1.64   **1.65**

Two families of compounds that are also derived from amino acids are the betalain pigments, *e.g.* betanidin **1.66**, and the glucosinolates. Betalamic acid, which is the central building block of the betalains, arises from the cleavage of dihydroxyphenylalanine. These pigments are typical constituents of the Caryophyllales and provide the red colour of the beetroots. The glucosinolates, *e.g.* sinigrin **1.67**, are formed by the decarboxylation of an amino acid and oxidation of the amine to the oxime, before the sulfur and a glucose unit are added to give the characteristic structural unit. The glucosinolates are typical constituents of the Brassicaceae where they have a role as natural pesticides. They are responsible for the bitter taste of members of this family. At high doses some are goitrogenic but at lower doses they exert a chemoprotective effect by inducing enzymes in the liver that detoxify harmful metabolites. Some of their beneficial properties can be attributed to the isothiocyanates that are formed by their decomposition.

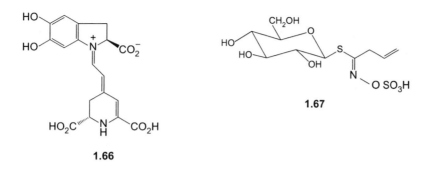

1.66

1.67

1.11 ALKALOIDS

Alkaloids are natural products which contain one or more basic nitrogen atoms. In contrast to the terpenes, there is no equivalent to the isoprene rule for alkaloids. The amino acids, principally ornithine, lysine, phenylalanine, tyrosine and tryptophan, are precursors of various families of alkaloids. The alkaloids are then classified in terms of their underlying heterocyclic skeleton or plant origin. We will come across specific alkaloids in the context of some vegetables. However, it is worth pointing out that many alkaloids were first studied because of their toxic properties. Some alkaloids have neuroactive properties and interact with receptors at nerve endings. Within their structures they can have fragments which resemble the neurotransmitters and facilitate their binding to these receptors. Not surprisingly, although alkaloids are found in fruit and vegetables, because of their toxic properties they are of less importance in edible crops than in the general plant kingdom. It is worth pointing out that some plants which are related to edible crops contain toxic alkaloids. For example, deadly nightshade (*Atropa belladonna*) is a member of the Solanaceae, a family which also includes the potato and tomato, whilst the pea and the lupins both belong to the Fabaceae family. There are also some alkaloids, such as the steroidal alkaloids, in which the nitrogen is introduced after the carbon skeleton has been formed. These compounds are found as toxic constituents of the inedible parts of potato and tomato plants.

1.12 VITAMINS

Vitamins are essential dietary factors for man which are required in small amounts (µg–mg per day) to maintain normal growth and development. We can obtain many of our vitamins either directly from fruit and vegetables, or by modifying constituents of plants. Many vitamins play an important role as co-enzymes in plant biochemistry, as well as in man. Vitamins were isolated mostly during the period 1910–1940 as a result of studies into the role of the constituents of foods in preventing deficiency diseases.

Some vitamins are fat soluble whilst others are water soluble. They have very diverse structures and biochemical functions. Some vitamins regulate aspects of mineral metabolism and cell differentiation or processes such as vision, whilst others function as co-enzymes in enzymatic catalysis or as antioxidants. As the individual vitamins were isolated, they were named using letters of the alphabet. Thus, vitamins A and D were lipid soluble, whilst the vitamin B group and vitamin C were water soluble. Although they were originally isolated from a particular plant or animal source, in practice most of the vitamins are quite widespread, albeit in small amounts.

Vitamin A **1.68** which is required for vision is obtained by metabolism of those carotenoids which contain a β-ionone ring, such as α- and β-carotene, and β-cryptoxanthin. These particular carotenoids are found in carrots, broccoli and spinach.

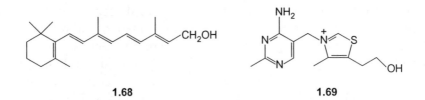

1.68   **1.69**

Vitamin B$_1$ (thiamin) **1.69** was discovered in a classical study of a deficiency disease. It was found that when the husks of rice were removed during rice polishing, a deficiency disease, beriberi, which affects the peripheral nervous system, occurred. Thiamine diphosphate is a co-factor in the oxidative decarboxylation of pyruvate and in a number of other important reactions of carbohydrate metabolism including transketolase. It is also a co-factor for the enzyme which catalyses the condensation of pyruvate and glyceraldehyde 3-phosphate to give 1-deoxy-D-xylulose-5-phosphate, from which many plant isoprenoids are biosynthesized. This vitamin is found in various parts of vegetables where carbohydrate metabolism is significant, such as in the grain and seeds, and in potatoes. Vitamin B$_2$ (riboflavin) **1.70** is a co-enzyme for the flavoproteins which mediate various oxidative processes. It is found in green beans and asparagus. Vitamin B$_3$ (niacin, nicotinic acid) was discovered in studies on the deficiency disease, pellagra. The amide forms part of the co-enzyme for the redox nicotinamide enzymes.

It is very widespread and is found in tomatoes, legumes, broccoli and carrots. Pantothenic acid, which was sometimes known as vitamin B_5, is the amide of D-pantoate and β-alanine. It is required to form co-enzyme A and is found in legumes. Vitamin B_5 (pyridoxal) **1.71** as the phosphate, is a co-factor for aspects of amino acid metabolism including trans-amination and decarboxylation. Various neurological conditions involving specific neurotransmitters derived from amino acids are caused by a deficiency of this vitamin. It is found in small amounts in many vegetables such as broccoli, carrots and potatoes. Folic acid (vitamin B_9, pteroylglutamic acid) is required for many one-carbon transfer reactions, such as the conversion of uracil to thymine in nucleic acid biosynthesis. As the name suggests it is isolated from leafy vegetables such as spinach. Unlike the other members of the vitamin B group, vitamin B_{12} is not found in vegetables, a problem for vegetarians.

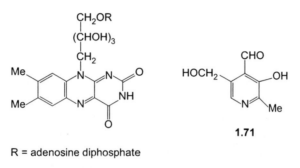

R = adenosine diphosphate

1.70

1.71

Vitamin C **1.72** is an important constituent of fruit and vegetables. Although it is formed from D-glucose, the configuration at C-5 indicates that it belongs to the L-series. The biosynthesis of L-ascorbic acid from D-glucose in plants such as strawberries, takes place *via* an oxidation to D-gluconic acid and 2,5-diketogluconic acid, followed by reduction to 2-keto-L-gulonic acid. The configuration at C-5 has been inverted in this step. Subsequent lactonization and enolization afford L-ascorbic acid. The amounts of vitamin C produced by plants vary quite widely with some plants, such as rosehips and blackcurrant, containing quite high quantites (200 mg per 100 g) and other plants, such as plums and apples, quite low amounts (*ca.* 10 mg per100 g).

Vitamin D and its precursors do not occur in plants but the fat-soluble vitamin E group (the tocopherols) **1.73** and vitamin K **1.74** are found in various seed oils and also in green leafy vegetables such as spinach. The tocopherols have antioxidant properties although it is not clear if this is their only function. The vitamin K series is a group of

2-methylnaphthoquinone derivatives which are important for blood coagulation and bone deposition.

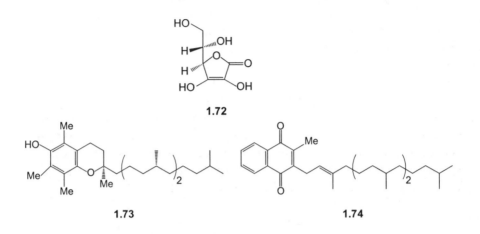

1.72

1.73   **1.74**

1.13 PLANT HORMONES

The enzyme systems responsible for plant growth and development are under hormonal control. The plant growth hormones are a structurally diverse group of substances that play a role in mediating various aspects of plant growth and development. These hormones do not just act independently but interact with each other. They are produced by plants in very small amounts, often at the level of μg per kg. Consequently their detection and much of our recent knowledge of their biological function has rested on highly sensitive instrumental methods of analysis, particularly gas chromatography linked to mass spectrometry. Some of these compounds, or their analogues, have been used in agriculture or horticulture to alter fruit ripening, to modify plant growth and to act as selective herbicides. The biosynthesis of these compounds and the genetics of plant hormone production have been the subject of intense study for over seventy years.

The first plant hormone to be detected was the auxin, indolyl-3-acetic acid **1.75**, which was identified in plants as a consequence of studies on the growth of plants towards light (phototropism). It is biosynthesized in the growing tips of plants and it induces pronounced plant growth and stimulates root growth and flowering. It has been claimed that a corn root will respond to as little as 10^{-12} g of indolyl-3-acetic acid. When a cutting of a plant is taken, the auxin that is formed in the growing tip is translocated downwards in the phloem to stimulate root formation. Synthetic analogues of indolyl-3-acetic acid, such as

2-naphthoxyacetic acid, are used as 'rooting hormones'. It has also been used in a fruit spray to enhance the setting of tomatoes, strawberries and grapes. 2,4-Dichlorophenoxyacetic acid (2,4-D) and its relatives, such as 2-methyl-4-chlorophenoxyacetic acid (MCPA), are present in 'hormone' weedkillers (see section 2.12)

The cytokinins, kinetin **1.76**, and the related zeatin and N^6-isopentenyladenine, are formed predominantly in young roots and stimulate cell division. Although there was evidence in 1913 that there were compounds that promoted cell division which were found in the phloem of plants and identified much later in coconut milk, it was not until the early 1960s that these compounds were fully characterized. To illustrate the importance of the influence of geometrical isomerism, the *trans* form of zeatin is 50 times more active than that of the *cis* form.

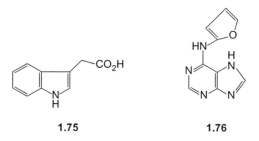

1.75   **1.76**

In contrast to the auxins which promote apical dominance, the cytokinins promote the development of side shoots. An alteration in the balance of these may account for the differences between the tall and shrubby varieties of plants such as the tomato. When a gardener wishes to grow a bushy plant, nipping out the growing tips removes the source of the auxin and allows the cytokinin to dominate. They also have an effect on the development of fruit and there is an application in conjunction with gibberellins A_4/A_7, to improve the development of some apple varieties. Cytokinin analogues can also extend the shelf life of cut flowers.

The gibberellins were originally isolated as the phytotoxic metabolites of a rice pathogen, *Gibberella fujikuroi* (*Fusarium moniliforme*) and now renamed *Fusarium fujikuroi*. The 'bakanae' (foolish seedlings) disease of rice, which involved an uncontrolled growth of the plants, was recognized as a fungal disease in 1898. In 1912, Sawada suggested that there was a growth stimulant produced by the fungus, and in 1926, Kurosawa showed that the culture filtrate from a fungal fermentation contained a substance which produced this effect. Crude material was isolated in 1938 by Yabuta and Sumiki. Much of the chemistry of the gibberellins was established with the fungal metabolite, gibberellic acid **1.40**, which was isolated at the Akers Laboratories of ICI in 1954. In the late 1950s, it was realised that the phytotoxic effect of the gibberellins involving excessive stem elongation, was actually an over-response to a normal hormonal effect. Gibberellins were then found firstly in 1958 by Mac-Millan in minute amounts (mg per kg) in beans and peas, and subsequently, in many other plant species. They have also been found in other fungi. Although gibberellins are present in minute amounts throughout the plant, the seeds and fruits have proved to be the best sources of material.

There are about 150 gibberellins that are known, each being designated as gibberellin A_n ($n = 1$–150). They have the same underlying unique tetracarbocyclic skeleton and differ from each other in their hydroxylation pattern. Recently, some gibberellins with an additional 9,15-cyclopropane ring have been discovered in apples. The gibberellins are all carboxylic acids. There are two series of gibberellins. One series retains the twenty carbon atoms of their diterpenoid precursor, whilst the other series has lost one carbon atom (C-20) and possesses a γ-lactone ring in its place as in **1.40**. The C_{19} compounds are formed from the C_{20} series.

The biosynthesis of the gibberellins was originally studied in the fungus and then in cell-free systems from developing seeds of *Pisum sativum* (peas), *Cucurbita maxima* (pumpkins) and *Marah macrocarpus* (wild cucumber). Isolation of the genes encoding the biosynthesis of the gibberellins from both fungi and plants has now revealed that there are some, originally unsuspected but fundamental, differences between the pathways in fungi and plants, particularly in the nature of the enzymes which mediate various oxidative steps. Thus fungi utilize cytochrome P_{450} systems for some hydroxylations, whilst plants use 2-oxoglutarate dioxygenases. However, the initial cyclization of the C_{20} diterpenoid precursor, geranylgeranyl diphosphate, to *ent*-kaurene **1.77**, followed by its oxidation at C-19 and then at C-7, and the ring contraction of **1.78** to the first gibberellin, gibberellin A_{12} aldehyde **1.79**, are similar but not identical. In plants, the aldehyde is oxidized to the acid **1.80** and then, depending on the species, hydroxylations may occur in different orders at C-3, C-13 or other centres and at C-20. The C-20 carbon atom appears to be lost at an aldehyde oxidation level **1.81** leading to gibberellins such as gibberellin A_1 **1.82**, gibberellin A_4 **1.83** or gibberellin A_{20} **1.84**. Thus in some plants, an early or a late C-3 or C-13 hydroxylation pathway may predominate. The deactivation and catabolism of the gibberellins begins with hydroxylation at C-2. The sequences that are operative may also depend on the maturity of the plant. Examination of

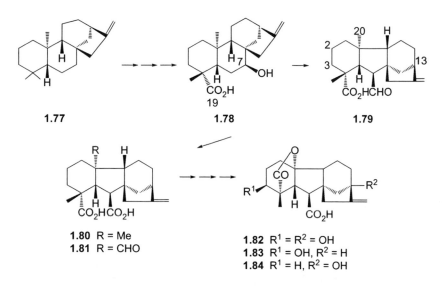

1.77 **1.78** **1.79**

1.80 R = Me
1.81 R = CHO

1.82 R¹ = R² = OH
1.83 R¹ = OH, R² = H
1.84 R¹ = H, R² = OH

any one plant by gas chromatography-mass spectrometry may reveal the presence of many gibberellins that are related to each other by a common biosynthetic sequence. The individual gibberellins that are present in particular fruit and vegetables are described in the relevant sections on these plants. The best known gibberellin is gibberellic acid (GA₃) **1.40** which is available commercially and is produced by fermentation.

Despite the large number of gibberellins that are known, biological activity is restricted to a few C_{19} gibberellins, such as gibberellin A_1, gibberellic acid, gibberellin A_4 and gibberellin A_7. Gibberellins have a number of biological effects. They stimulate cell division and cell elongation including increasing stem length. They stimulate flowering and delay senescence of leaves and some fruit. They break seed dormancy and induce the formation of α-amylase, particularly in germinating cereal grains, and they cause parthenocarpic (seedless) fruit to develop, as in seedless grapes. In the developing seed, mobilization of gibberellins leads to the formation of α-amylase and the breakdown of starch reserves into sugars. In growing plants, they interact with DELLA proteins leading to their modification. DELLA proteins, named after a sequence of their constituent amino acids, have a regulatory effect on plant growth. Their degradation leads to the unrestricted growth characteristic of the 'bakanae' disease.

Since gibberellins influence many aspects of plant growth and development, they have a number of commercial uses. For example, the enhancement of the action of α-amylase in releasing sugars from starch is used in malting barley in beer manufacture. A gibberellin concentration of about 0.5 mg per kg of barley is required. Gibberellins are also used in the treatment of citrus fruit, inhibiting senescence, maintaining a healthy skin and reducing the ease of attack by pests. A combination of gibberellin A_4 and 6-benzylaminopurine has been marketed to stop excessive russeting in fruit. Another commercial application is in a spray for improving the size of seedless grapes. Some synthetic plant growth regulators, such as paclobutrazole (Bonzi®) and chlorocholine chloride (CCC), are inhibitors of gibberellin biosynthesis. Thus paclobutrazole inhibits the oxidation of *ent*-kaurene at C-19. Gibberellin biosynthesis is blocked in some genetically dwarfed plants.

A group of plant hormones which have been discovered more recently are the brassinosteroids, exemplified by brassinolide **1.85**. These steroid hormones, of which there are some sixty known compounds, were originally discovered in 1979 in the pollen of oil seed rape, *Brassica napus*, as compounds which promote cell division and expansion. They are now known to be widespread throughout the plant kingdom in very low concentrations, typically 10^{-1} nmol per g. For example, they have been found in the pollen of the sunflower (*Helianthus annuus*) and in peas (*Pisum sativum*), beans (*Phaseolus vulgaris*), and radish (*Raphanus sativus*). Since brassinosteroids occur in minute amounts, much of their biological activity has been established with synthetic material prepared from readily available plant sterols. The brassinosteroids appear to stimulate a number of aspects of root and stem growth, acting synergistically with indolylacetic acid **1.75**. They influence a plant under conditions of stress, conferring heat and cold tolerance and resistance to drought. This anti-stress effect may have potential applications in crop production.

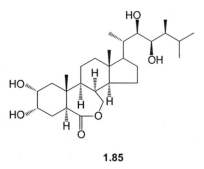

1.85

Whereas the plant hormones that have been described so far stimulate plant growth, abscisic acid **1.38** modulates several aspects of dormancy. Indeed its early name was 'dormin'. Abscisic acid is formed in the autumn and brings about leaf and fruit fall. It is also formed by plants under water stress and protects the plant against excessive water loss.

It can even lead to the loss of leaves in the height of summer. In many ways, abscisic acid is an antagonist of the plant growth hormones. It appears to induce seeds to synthesize storage proteins and inhibit the action of gibberellins on α-amylase. Although abscisic acid has a sesquiterpenoid structure, it may be formed in plants by the biodegradation of the carotenoid, violaxanthin *via* a xanthoxin. The isoprene units of the carotenoids that are biosynthesized in the chloroplasts are derived by the deoxyxylulose pathway. Abscisic acid in plants is formed by this pathway. Abscisic acid is also formed by some fungi that are plant pathogens, such as *Cercospora rosicola*. These organisms use the mevalonate pathway to form abscisic acid directly from a sesquiterpenoid precursor, farnesyl pyrophosphate. Fungal and plant abscisic acids appear to be biosynthesized by different routes.

Ethylene is an unusual hormone and plays an important role in the ripening of fruit. It is formed in ripe fruit and induces ripening in others. Green tomatoes and other climacteric fruits may be ripened under a sheet of paper to trap the ethylene, provided that there is a ripe tomato or even a ripe banana present to provide the ethylene. Ethylene is biosynthesized from the amino acid, 1-aminocyclopropanecarboxylic acid **1.86**. The production of ethylene can affect the storage of fruit. Apples that are only producing a low concentration of ethylene (less than 0.1 ppm) will store quite well in a cold area, whilst those that are producing more ethylene do not store well. The inhibition of ethylene biosynthesis by compounds such as 2-amino-4-aminoethoxy-*trans*-3-butenoic acid (aminoethoxyvinylglycine, AVG) or the function of ethylene by the gas, 1-methylcyclopropene (MCP), delays senescence and extends the storage life of fruit. Some of the 'out-of-season' fruit on the supermarket shelves has been preserved by the action of MCP.

A number of other compounds play a signalling role in plant chemistry. Jasmonic acid **1.20** is produced under stress and in low concentrations. It and its more volatile relative, jasmone, stimulate the defensive system of plants. Application of the methyl ester of jasmonic acid induces the formation of tubers in potatoes. It is related to the natural tuberonic acid. Jasmonic acid also induces ethylene synthesis and it can inhibit seed germination. The cyclopentenone, *cis*-jasmone is produced by decarboxylation of jasmonic acid and is a well-established component of plant volatiles. It is released when a plant is damaged. Whilst *cis*-jasmone has a repellant effect on some insects such as aphids, it acts as an attractant to their predators such as ladybirds. Along with other leaf volatiles including *cis*-hex-3-enal and methyl salicylate, *cis*-jasmone also has a signalling effect on neighbouring plants priming them against impending herbivore attack. In Brassicas, such as the turnip, methyl salicylate and jasmonic acid have been shown to stimulate the formation of glucosinolates. Salicylic acid **1.87** also influences the growth of flowers, buds and roots. When it is added to the water in a vase containing cut flowers, it can delay their withering.

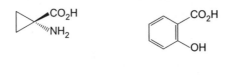

1.86   1.87

Thus there are a variety of natural products of widespread occurrence in plants that play an important role in general plant biochemistry. These natural products contribute to the beneficial and organoleptic properties of fruit and vegetables.

CHAPTER 2

Chemistry and the Growing Environment

2.1 THE SOIL

The contents of the soil involve an interaction between inorganic and organic chemistry in facilitating the growth of fruit and vegetables. Much of what can grow above ground is determined by what is available to the plant below ground level. The soil is a mixture of the products of disintegration of the Earth's outer crust from the weathered bedrock and the material deposited by glaciation, rivers and from the atmosphere, together with the decaying remains of plants and animals and a wealth of living organisms. Over the centuries the activities of man have modified many soils. The soil, particularly the twenty to thirty centimetres of topsoil, is a dynamic growth medium changing with time both in the short term through the seasons, and in the longer term over many years. The depth of this topsoil and its relationship to the subsoil and any underlying clay, chalk, sand or granite can have an impact on the drainage and on the growth of deep-rooted shrubs and trees.

In the previous chapter we have seen the role that various macromolecules and natural products formed by the plant play in the development of fruit and vegetables. The macroscopic properties, the mineral content and the organic humus in the soil are important in determining the health and development of the root systems of plants which in turn affect the overall chemistry of the plant. The water, various metal ions, phosphate- and nitrogen-containing compounds required for the

Chemistry in the Kitchen Garden
By James R. Hanson
© James R. Hanson 2011
Published by the Royal Society of Chemistry, www.rsc.org

biosynthesis of natural products are transported from the roots in the xylem to the upper plants of the plant. In turn sugars and other compounds are returned to the roots in the phloem. Large parts of the structure of the plant are concerned with the efficient transport of these from the soil *via* the roots to sites of biosynthesis within the plant and then from these sites to target organs and sites of storage.

2.2   THE MINERAL STRUCTURE OF THE SOIL

Soils are classified in terms of their particle size. An approximate division is that particles which are greater than 2 mm in diameter form gravel and stones. Those particles with a diameter between 0.1 and 2 mm are sand, whilst those that are between 0.005 and 0.1 mm are silt and those that are less than 0.005 mm are clay. The pores within the soil are determined by the particle size. The larger transmission pores (> 50 μm in diameter) allow water and root penetration. These larger pores are associated with lighter sandy soils and, apart from roots, they also allow fungal mycelium to spread within these soils. Narrower pores (0.2–50 μm in diameter) store water for plant use but the smallest pores (< 0.2 μm in diameter) retain water in a hydrogen-bonded network and produce a heavier soil. The pores within the soil not only affect water and thus metal ion availability but also determine aeration and the rate of temperature change with the seasons. Many of the factors which affect chromatography in the laboratory are replicated in the soil – compare the ease with which solvents penetrate the different grades of silica.

Whereas sand is mainly quartz or silica, the clays are the weathered products of silicate minerals. The clays contain hydrated aluminium silicates with up to 50% SiO_2 and 30–40% Al_2O_3. There are smaller amounts of Fe_2O_3, TiO_2 and hydrated CaO and MgO. These variations produce the many varieties of clay that are found in this country. Clays contain various distinct minerals, such as kaolinite and montmorillonite, some of which also find use in the laboratory as catalysts or catalyst supports.

The water content of clays means that these small particles bind together and are classified as 'heavy'. In the spring they take longer to warm up and when they dry out they form a crust and there is contraction leading to cracks in a lawn, or even worse effects on house foundations. These crusts and cracks also have an obvious effect on the local distribution of water to plants in dry periods and on the consequences of sudden summer downpours. The importance of an autumn dig to allow the winter frost to penetrate and cause this interstitial water to freeze and break up the clods is well-known. However, the expansion in forming ice can also damage the roots of plants. Apart from its effect on growing weeds, spring hoeing can reduce the formation of cracks and enhance the penetration of rainfall in summer. The role of the pores in the soil is crucial to managing the availability of water to fruit and vegetables which, with a high metabolic turnover, can be quite thirsty plants. The ease with which mineral resources can be leached by roots from the soil is proportional to the surface area to volume ratio of the particles. Some clays have a colloidal fraction which can be coagulated by the addition of lime.

The underlying materials that contribute to the soil may be igneous rocks, such as granite, metamorphic rocks, such as slates, or sedimentary rock, such as chalk or limestone. As the molten magma of the earth crystallized to form the igneous rocks, the mineral content changed from olivine, a magnesium silicate, and various feldspars containing aluminium silicates together with sodium, potassium, calcium and barium oxides to the formation of iron-rich minerals, such as pyrites, magnetite, chromite and ilmenite. Finally rocks with a high silica content were formed. Rocks with less than 45–55% silica are known as 'basic' rocks, whilst those with greater than 65% silica which often contain free quartz are 'acidic' igneous rocks. On weathering, igneous rocks give soils containing quartz and mica. However, those sedimentary rocks which were formed from the skeletons of marine organisms give calcite or dolomite (a mixture of magnesium and calcium carbonate).

The structure of the silicates determines many of the properties of the resultant rock and hence the derived soil. The dominant feature of silicate minerals is the tetrahedral SiO_4^{4-} unit. In simple olivine minerals, this is associated with divalent cations such as Mg^{2+}, Fe^{2+} and Ca^{2+}. However, if there are shared oxygens between SiO_4 units, a silicate chain, sheet or three-dimensional network can develop. These can bind metal ions. In some instances there is also replacement of silicon by aluminium. The hydroxides of aluminium and magnesium can also form sheet structures.

Clay minerals, such as kaolin and montmorillonite, contain layers of tetrahedra and sheets, sometimes with cations or water molecules in the interlayer spacing. This gives rise to variable cation exchange properties and to changes in shape between hydrated and dried materials.

In the UK, the parent material of the soil has often been moved some distance by the extensive glaciation that once affected this country. This soil movement alters the nature of the soil and its stone and flint content. Water erosion has also had a significant impact on the minerals in the soil, particularly affecting the alkali and alkaline earth metal content.

Human factors have also led to changes in the soil and these are discussed later.

2.3   THE ORGANIC CONTENT OF THE SOIL

The degradation of plant material leads to the accumulation of humus in the soil. Amongst other roles, this organic matter serves to bind the inorganic particles together and act as a reservoir of particular nutrients. The black colloidal powder which is known as 'humus', is divided into alkali soluble and insoluble fractions, the humic acids and humin. The humic acids are high molecular weight polymeric polyhydroxy phenolic acids that are derived from, for example, caffeic acid **2.1** and 3,4-dihydroxybenzoic acid **2.2**. They have a variable number of polysaccharide and protein units attached to them. A typical analysis might be 50–60% carbon, 4.5–5.8% hydrogen, 2–4% nitrogen and 33–35% oxygen. Whilst the cellulose of plant material decomposes relatively rapidly, the aromatic components of the lignin are more resistant and they provide the basis of the humic acids.

The presence of the hydroxyl and carboxyl groups provide ample opportunity for complexing metal ions, particularly iron, but also lead, copper, nickel and zinc. It has been suggested that humic acids may hold up to 10% of their weight as metal ions. The combination of cation exchange sites in clay minerals and the metal-binding properties of the soil organic matter, leads to the concept of a cation-exchange capacity of the soil. This determines the availability of some essential nutrients including the ammonium ion, for the developing plant. The humus within the soil can play an important part in holding together soils which crumble too easily. Poor sandy soils in coastal regions are often improved by digging in kelp to enhance the mineral content, particularly potash and the organic content of the soil.

The lower molecular weight fulvic acids are more water soluble and they sometimes give a brownish colour to surface and marshy water. A normal concentration in surface water might be 1–5 ppm. A typical analysis of the fulvic acids gives a composition of 42–50% carbon, 4–6% hydrogen, 1–2% nitrogen and 45–47% oxygen. Compared to the humic acids, there is a higher oxygen content. The fulvic acids not only complex metal ions but may also bind phosphate units. The binding of phosphate by organic matter in the soil can seriously affect the availability of phosphate for plant growth. Inositol penta- and hexaphosphates, such as phytic acid **2.3**, can account for a substantial proportion of the total soil phosphorus. These inositol phosphates are only slowly biodegraded to release phosphate.

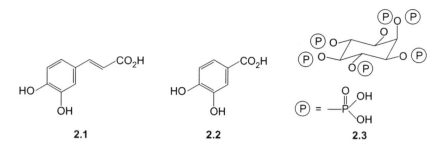

2.1	2.2	2.3

As the roots of a plant penetrate the soil, they produce a poly-saccharide exudate which facilitates the uptake of water and metal ions by the roots. The development of this layer is sometimes apparent around germinating seeds. It will hold soil to the root system and it plays an important part at the interface between the soil and the plant. Care has to be taken not to disturb this in planting out seedlings. Micro-organisms in the soil also produce a polysaccharide exudate which plays a role in the transport of nutrients. Many plants have mycorrhizal fungi associated with them which facilitate this mobilization of nutrients. When transplanting fruit bushes or trees, the plant can be helped to become established by adding a handful of a mycorrhizal preparation. In the competitive microbiological world of the soil, this can help to favour the beneficial rather than the phytopathogenic organisms.

The primary alcohols of the sugars in the polysaccharides that con-stitute the pectin of fruit are oxidized to a carboxylic acid. Biode-gradation of this material in the soil gives a polyuronic acid which can also bind the clay particles together and form salts with metal ions.

2.4 NUTRIENTS FROM THE SOIL

Nitrogen in the form of soluble ammonium salts, phosphorus as phos-phates, sulfur as sulfate, together with potassium, calcium and magne-sium cations are regarded as macronutrients for plants. The elements sodium, iron, manganese, copper, zinc, molybdenum, nickel, cobalt, boron, selenium and chlorine in various combinations form essential micronutrients, whilst the availability of other elements, such as silicon and vanadium, are beneficial. However, it is worth noting that an excess of some of the micronutrients can be toxic.

Compounds containing nitrogen, such as the amino acids and various heterocyclic bases, are ubiquitous. Pale chlorotic leaves arising from a lack of chlorophyll and feeble plant growth are a consequence of nitrogen deficiency. Nitrogen-fixing bacteria such as *Azotobacter* and the *Rhizobium* species associated with root nodules on legumes, convert dinitrogen to ammonia. On the other hand, *Nitrosomonas* and *Nitrobacter* species convert ammonia to nitrites and nitrates. This bacterial nitrification can then represent a loss of nitrogen. By incorporating nitrogen into their structure, soil bacteria and fungi immobilize nitrogen. On the other hand, by degrading plant material, nitrogen can be released by a process known as 'mineralization'.

Phosphates play an important part in energy transfer as ATP, as constituents of the nucleic acids and in various biosynthetic processes. Phosphate is obtained by the degradation of the phytic acid **2.3** and it is released from the fulvic acid, as well as from mineral sources. Sulfur is a component of the amino acid, cysteine, and of a number of more specialized natural products, particularly those typical of the Alliums and the Brassicaceae. When growing these plants it is sometimes advantageous to add sulfur or a sulfate to the soil.

Metal ions, such as magnesium and iron, play an important role in the co-enzymes that mediate photosynthesis and oxidative processes. Apart from their role as a constituent of chlorophyll, magnesium ions play a part in many enzymatic reactions involving ATP. A magnesium or an iron deficiency may be manifest in chlorotic leaves. A magnesium deficiency can be corrected by adding a small amount of magnesium sulfate (Epsom salts).

Iron is essential for the growth of plants because of the ability of its biological complexes to co-ordinate and activate oxygen and then deliver it to an organic substrate in various oxidative steps. Its redox chemistry, based on the Fe(II) $\Leftrightarrow$ Fe(III) $\Leftrightarrow$ Fe(IV) interchange, is essential for the electron transport involved in these metabolic processes. Despite the abundance of iron in the environment, iron(III) is poorly soluble particularly in alkaline soils. Consequently, plants and microorganisms have evolved various strategies such as the use of siderophores to scavenge their iron requirements from the soil. Siderophores are molecules that have the ability to complex and transport iron(III). Typically they contain hydroxamate, catechol or α-amino or α-hydroxy carboxylate residues which are often incorporated into hexadentate structures that form octahedral complexes with iron(III). They often have a higher affinity for iron(III) than iron(II), and are therefore less likely to complex other divalent cations such as zinc(II) or copper(II). An example is vicibactin which is produced by *Rhizobium leguminosarum*, a bacterium found in root nodules on peas. This is a cyclic trimer of N^2-acetyl-N^5-hydroxyornithine and (R)-3-hydroxybutanoic acid arranged alternately with alternating ester and peptide bonds. An iron deficiency can be rectified by the use of the soluble iron–ethylenediaminetetraacetic acid (EDTA) complex, Sequestrine®.

Other metal ions, such as those of potassium, calcium, manganese, molybdenum, zinc and copper, play a role in various cellular processes involving water transport, cell wall formation, second messenger action, electron transport and holding proteins in particular conformations. For example, zinc is involved in complexation to histidine side chains in carbonic anhydrase and alcohol dehydrogenase, whilst copper is involved in electron transport. Their deficiency, particularly in heavily cultivated soils, can have a serious effect on crop yields.

Vegetables provide an important source of minerals in our diet. The concentration of minerals in vegetables depends on the role that the mineral plays in the development of the plant and are often in the order of mg per 100 g fresh weight. Typical concentrations (per 100 g) are calcium (10–50 mg), iron (0.5–3 mg), zinc (0.1–1 mg), magnesium (12–33 mg), phosphorus (50–100 mg), potassium (170–400 mg) and sodium (3–6 mg). For example, in potatoes, potassium may be present in relatively large amounts (up to 400 mg per 100 g), whilst sodium is present in much smaller amounts. Although aluminium is a major constituent of soil (up to 8%), it does not make a major contribution to plant development and it can be harmful to man, producing neurotoxic effects and damage to bones. The accumulation of toxic metals and the role of fertilizers are discussed later (sections 2.6 and 2.9).

2.5 THE EFFECT OF PH

The availability of metal ions for uptake by plants is particularly dependent on the pH of the soil. This is an extremely important parameter in determining the availability of nutrients for the developing root system. In an alkaline soil, iron, zinc and manganese are held as their less soluble hydroxides, whilst they are more readily available as salts from an acidic soil. On the other hand, a low soil pH, typically below 5.0, can have an adverse effect on the growth of plants producing aluminium and manganese toxicity and a deficiency of molybdenum, calcium and magnesium. Nodules on legumes may fail to develop whilst fungi which grow in acidic soils may develop. These factors influence the growth of ericaceous plants and popular fruits, such as blueberries, which require an acidic soil. An optimum pH for the growth of many plants is 6.5. Many bacteria grow at a pH nearer to 6.5 whilst fungi prefer a more acidic soil, a feature which can have an effect on the diseases of fruits and vegetables.

2.6 BROWN FIELD SITES

The previous history of the site of a vegetable plot or allotment is worth exploring. The soil may be affected by the previous use of the land. An aim of the Allotment Acts of the late 19th and early 20th century was to bring more land into use as urban allotments. One source of land for this purpose was railway land, some of which had been used for depositing soil from cuttings and other earth movements. More recently, former factory sites and even former railway lines have become sites for housing, their gardens and allotments. Some allotments are adjacent to sewage works or on land which has been restored from former mine workings or waste tips, and may also be subject to 'run-off' from them.

Not all that long ago, the common practice for disposing of chemical waste was to bury it. Sewage sludge may contain elevated levels of cadmium, zinc, copper, chromium and lead. Even former horticultural land may contain higher than normal levels of copper, arsenic or lead. Lead arsenite was used up to fifty years ago as an insecticide spray in orchards, whilst copper products were used as fungicides. Copper chrome arsenate has been used as a wood preservative. Cuprinol® is a copper naphthenate. Studies have shown that these elements can accumulate in vegetables such as cabbages, onions, beetroot, lettuce and radishes. There have been cases of allotments in the London area which had to be closed as a result of high levels of lead. Buried iron work, paint and electrical components can make a significant contribution to soil trace metal composition. Problems of cadmium poisoning have been associated with vegetables grown on former mine workings.

Selenium provides an interesting case because selenium compounds can find their way into soil from a variety of sources including the modern electronics industry. Selenium is an essential element for humans as a micronutrient but at higher levels it can be toxic. Its availability for plants such as Alliums is pH dependent. Selenates are mobile in normal and alkaline soils. The pH of the soil is also an important parameter which affects the bioavailability of toxic metals.

The Environment Agency is producing a series of soil guideline values for the concentration of particular elements, such as selenium, according to land use. The UK Soil and Herbage Survey contains further information concerning contaminants in soils across the United Kingdom.

2.7   MICROBIAL INTERACTIONS WITHIN THE SOIL

The soil is a dynamic biological medium in which bacteria, fungi, nematodes and various insects flourish in competition with the developing plant. Evidence is growing for the role of volatile compounds in mediating the below ground interactions between plant roots and soil organisms. The roots of many plants have mycorrhizal fungi associated with them in a symbiotic relationship. Whilst these fungi draw their sugars from their host plant, they mobilize nutrients from the soil to the benefit of the plant. They are particularly important in releasing phosphate, accumulating nitrogen in a readily available form for higher plants and in mobilizing iron. The beneficial effect of fungi in releasing nutrients can be seen in the enhanced growth of grass just within a 'fairyring' of the fruiting bodies of a fungus such as *Marasmius oreades* in the lawn. Further inside the ring, the grass may then die as a result of the action of other fungal metabolites including polyacetylenes such as agrocybin. Another 'fairy-ring' organism, *Lepista sordida*, has been shown to produce both a plant growth stimulating substance, 2-azahypoxanthine **2.4**, and a plant growth regulator, imidazole-4-carboxamide **2.5**.

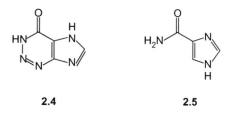

2.4	2.5

Many bushes and trees will not flourish without their associated mycorrhizal organisms. Hence in transplanting bushes, it is helpful to ensure that the soil associated with the plant is moved as well. The introduction of mycorrhizal organisms into 'brown field' sites may also facilitate the release of bound nutrients particularly phosphate and iron for the benefit of the plant. The roots of plants have also been found to stimulate the germination of mycorrhizal fungi. Recently this activity has been associated with deoxystrigol derivatives which can also affect the development of parasitic weeds.

The soil is home to root-infecting fungi which can have a destructive effect on fruit and vegetables. The dissemination of the spores of phytopathogenic fungi can take place through the air, the soil and by insect and animal vectors. The secateurs used to prune fruit bushes can transmit plant diseases. The spread of fungi in the soil depends on water and warmth as well as insect vectors. The 'damping-off' of young seedlings by *Pythium ultimum* or *Rhizoctonia solani* by over-zealous watering exemplifies this.

Many phytopathogenic fungi are host-specific and their implication in plant diseases will be discussed in later chapters on the individual crops.

There are, however, some general features relating to the interaction between a fungus and a plant. These interactions involve a combination of chemical and enzymatic processes. The initial step is recognition between the fungus and the plant, and we have already noted the influence of root exudates on the germination of fungal spores. The fungus must then gain entry to the plant either through a wound or by the action of cellulolytic enzymes. The wound in the root may arise from attack by an insect, nematode, wireworm, larva or beetle or from mechanical damage such as wind, which can produce abrasions against a stone. The invading fungus then produces degradative enzymes that cause necrotic lesions, as well as low molecular weight toxins which damage the plant. The phytotoxins may be translocated to the leaves in order to exert their effect, or the lesions and fungal polysaccharide exudates may affect the water and nutrient supply to the leaves, producing wilts. The action of various *Fusarium* species illustrates these steps. The cellular targets of the phytotoxins may be the inhibition of specific enzymatic processes, such as electron transport, or an alteration to the permeability of cellular membranes.

Many vegetable cultivars have been bred containing protective antifungal agents (phytoanticipins) or which respond to pathogen attack by producing phytoalexins. Phytoalexin production can be induced in healthy tissue by volatile compounds, such as methyl salicylate or methyl jasmonate, produced by neighbouring diseased tissue. Phytoalexins have fairly general antifungal activity. Virulent strains of specific plant pathogens have developed the ability to metabolize and detoxify phytoalexins. Some examples will be discussed later.

Soil-borne fungi do not act in isolation and some produce antifungal and antibacterial agents to establish their ecological niche. For example, very common *Trichoderma* species, such as *T. harzianum*, although not plant pathogens, nevertheless behave as competitive soil microorganisms. They exert their dominance over other microorganisms in three ways. Firstly, *T. harzianum* produces a volatile metabolite, 6-n-pentylpyrone **2.6**, which permeates through the surrounding soil and inhibits the germination of other fungi. Secondly, a group of antifungal agents, such as harzianopyridone **2.7** and gliotoxin **2.8**, are formed which are close-contact antifungal agents. The third stage involves the production of extra-cellular enzymes that destroy the fungal cell wall of other competitive organisms. Not surprisingly, *Trichoderma* species have been used as biocontrol agents (Rootshield®, Plantshield® or Trichodex®), whilst synthetic 6-n-pentylpyrone has been used as a soil fumigant and to prevent the microbial spoilage of fruit in storage.

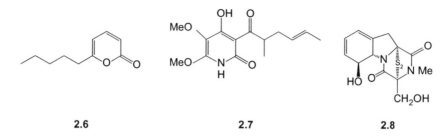

2.6　　　　　　　　　**2.7**　　　　　　　　　**2.8**

Many soil bacteria are involved in the biodegradation of plant material and cause plant diseases with symptoms that are similar to those of fungal infections. However, bacteria reproduce at a greater rate and the consequences appear more quickly. Other bacteria have a beneficial role. The free-living *Azotobacter* species and the symbiotic *Rhizobium* species found in the root nodules of legumes, such as peas and beans, carry out nitrogen fixation converting atmospheric nitrogen into ammonia.

Bacteria produce a number of volatile metabolites whose presence can be smelt and are indicative of the condition of the soil or a compost heap. Reductive anaerobic, often compacted, conditions lead to the formation of low molecular weight amines and hydrogen sulfide. The 'earthy smell' of newly dug soil is due to geosmin **2.9**. This norsesquiterpene is produced by soil *Streptomycetes*. The presence of geosmin and another bacterial metabolite, 2-methylborneol **2.10**, can be a problem in water supplies because the human nose can detect them at very low concentrations.

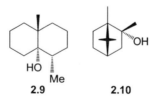

2.9　　　　　　**2.10**

2.8 CROP ROTATION

The exhaustion of available trace nutrients together with the potential build up of pests and diseases within the soil, leads to the requirement for crop rotation. A typical three-year rotation might be root crops in year one, followed by Brassicas and finally legumes. Where the structure of the garden and its microclimate (*e.g.* wind shelter) makes this impracticable, it is nevertheless important to at least try to move the soil around to minimize the problems. The microclimate of even a small kitchen garden is important to consider. There can be substantial and surprising variations in the amount of rainfall actually reaching the soil between exposed areas and those that are in the lee of a hedge or adjacent to a house. There can also be marked variations in the exposure to sunshine, temperature and destructive wind eddies around a house. One strategy is to use buried pots or containers whose contents can be changed. Wherever pots are used, it is wise to give them a rinse with bleach between years to clean them. Another strategy is inter-cropping, avoiding the monoculture which allows diseases to be passed from one plant to another and to use cover crops and companion planting, such as marigolds which have insect deterrent properties. Marigolds produce phenylheptatriyne and a series of thiophenes, such as α-terthienyl, arising from the addition of sulfur to the alkynes which have this property. Chives are another plant with deterrent properties. Since insect pests may be attracted to their host plants by specific volatiles from a damaged plant, a case can be made for some mixed cropping to reduce the impact of one damaged plant on adjacent healthy plants of the same species.

2.9   FERTILIZERS

Deficiencies of essential elements in the soil can lead to problems of plant growth. Fertilizers containing ammonium nitrate or ammonium sulfate, calcium phosphate and potassium chloride or nitrate are used to rectify these deficiencies. It is worth noting that when ammonium nitrate is used as a nitrogen source, the ammonium ion is assimilated first and the soil becomes more acidic. Urea is sometimes used as a nitrogen source.

The composition of fertilizers is not always specified directly as percentage nitrogen, phosphorus or potassium but as percentage nitrogen, phosphorus pentoxide and potassium oxide, although no fertilizer would contain P_2O_5 and K_2O! This reflects the analytical methods that were used to quantify these elements. Hence a typical 'Growmore®' fertilizer that is described as NPK 7:7:7 has a total nitrogen content of 7%, a phosphorus content of 3% and a potassium content of 5.8%, whilst a 4:4.5:8 'Tomorite®' with added magnesium ions contains 4% nitrogen, 2% phosphorus, 6.6% potassium and 0.018% magnesium.

Localized high concentrations of fertilizers can be toxic to young plants, a problem sometimes encountered with chicken manure pellets which contain a high concentration of urea. Care needs to be taken to water a fertilizer into the soil so that it becomes more evenly distributed and partially bound to organic matter in the soil and thus converted to a slow-release form. This can be a problem in potting-on and planting-out seedlings, particularly where the aim is to develop a good root system. It is also unwise to add lime to the soil at the same time as fertilizers, as this can lead to local alkaline conditions, the loss of ammonia and the conversion of any added trace metals to insoluble hydroxides.

An aqueous extract of weeds, such as the stinging nettle, can be a useful source of nitrogen. Extracts of comfrey (*Symphytum officinale*) or of a hybrid Russian comfrey (*Symphytum x uplandicum*) have been recommended as an 'organic' fertilizer. Comfrey is a deep-rooted plant which grows in damp soil and reaches a height of up to a metre. However, once it is established in the garden, it is difficult to get rid of it since it will develop from traces of root cutting. The deep roots bring minerals to the plant on the surface. The leaves and stems break down and can be fermented to give a dark liquid which when diluted makes a fertilizer that is high in nitrogen and potassium. The leaves contain a relatively high concentration, up to 4%, of allantoin (5-ureidohydantoin; $C_4H_6N_4O_3$) **2.11** and a similar amount of asparagine. Both are good sources of nitrogen. Allantoin is an oxidation product of uric acid, part of purine metabolism, and readily undergoes further decomposition *via* allantoic acid (diureidoacetic acid) **2.12** to release ammonia. Allantoin is present in a number of cosmetics and promotes cell proliferation and wound healing. This may account for some of the beneficial properties of comfrey. However, the leaves also contain some pyrrolizidine alkaloids such as symphitine and echinidine, which are hepatotoxic and care must be taken in using comfrey as a herbal tea. Wood ash is a good source of potash and, because the ash can be alkaline, it may favour the growth of plants such as the Brassicas.

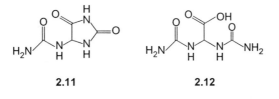

2.11   **2.12**

2.10 COMPOST

Although compost can be a useful source of nutrients, it is also of value in improving the general quality of the soil, its structure and water capacity. The John Innes composts are made up of a loam base (7 parts) with a peat substitute (3 parts) and sand (1 part), together with added fertilizer. Depending on the intended use, this may be 'hoof and horn' from the slaughter house to provide nitrogen and some minerals, calcium phosphate and potassium sulfate.

The use of the chemically much more complex manures and compost as a source of nutrients involves a balance between the requirements of the bacteria that create the compost and the plant material that is being degraded. This is particularly true in terms of available nitrogen. Bacteria have a lower carbon to nitrogen ratio (approximately 4 : 1) compared with the plant material that is being returned to the soil *via* the compost heap (approximately 20–30: 1). Consequently, a compost heap with actively growing bacteria can become depleted in nitrogen as the bacteria immobilize it in their peptides and proteins. Indeed an effective compost heap may need feeding with a nitrogen source such as ammonium sulfamate or urea as a compost accelerator. The home-made compost which is spread on the soil, whilst it may be beneficial for other reasons, can at first deplete the soil of nitrogen. For a successful compost heap, it is important to arrange the structure to encourage the growth of aerobic thermophilic bacteria. The initial degradation may involve mesophilic organisms, but in order to achieve a temperature of *ca.* 60 °C which destroys the seeds of common weeds, thermophilic organisms need to be encouraged. In order to spread these organisms through the compost heap, they need adequate moisture and a high plant surface area. Turning a compost heap facilitates this, as well as increasing the aeration. Too great a volume of grass cuttings in a compost heap can become compacted and anaerobic. When turning a compost heap, it is worth smelling it. A smell of ammonia is a sign of anaerobic conditions in contrast to the more fruity smell of esters arising from aerobic degradation. It is unwise to add meat to a compost heap as a source of nitrogen because the smell of putrescine can attract vermin.

Worms can digest material in the compost heap but the temperature is normally too high for the ordinary garden worm to survive. Whereas the carbohydrates of plant material may be degraded within six months, the lignin takes much longer and woody material may remain for some time. The phenols present in conifer cuttings have sufficient antimicrobial action to significantly restrict their biodegradation.

Various plants have been recommended for use as green manures. These cover crops are particularly useful in the garden or allotment when an early crop, such as potatoes, has been harvested and vacant soil needs to recover and to be protected from weeds. The green manure when it is dug into the soil, not only increases the organic matter in the soil improving water retention, but it also supplies nutrients. Typical green manure crops which can be sown in August or September are quick growing clovers (*Trifolium sp.*), winter tares (*Vicia sativa*) or winter field beans (*Vicia fabae*). These are members of the Leguminosae (Fabiaceae) which have nitrogen-fixing bacteria associated with them and thus they will increase the nitrogen content of the soil. Another crop are the rapidly growing mustards (*Sinapis alba* or *Brassica juncea*). These are members of the Brassicaceae which produce glucosinolates. These compounds decompose in the soil to release isothiocyanates which act as biofumigants, suppressing soil pests and diseases. Another effect is their inhibition of bacterial nitrification, a process by which ammonium ions are oxidized to nitrite or nitrate. The inhibition of this process can lead to an increase in the availability of nitrogen from the soil. This also suggests a use for the cabbage water from cooking which may contain glucosinolate degradation products.

The addition of plant material such as grass cuttings to the surface of the soil as a mulch can have a beneficial effect, not just in terms of water retention but also in preventing light from reaching the seeds of weeds and stimulating their germination.

2.11 WEEDS

From the gardener's point of view, a weed is a plant growing in the wrong place, but from the point of view of the weed, it is a place where the weed is succeeding in a competitive environment! Some weeds exert their dominance by physical means in outgrowing other plants and competing effectively for moisture and nutrients. Others exert a noxious effect by producing allelochemicals that inhibit the germination and development of other plant species.

Before clearing the ground either on an allotment or in a potential vegetable patch in the garden, identification of the common weeds that are growing there can give an idea of the condition of the soil, its pH, the drainage, depth and fertility. Although many weeds are ubiquitous, some favour an acidic or an alkaline soil whilst others require a damp, deep and fertile soil.

The presence of weeds such as the stinging nettle (*Urtica dioica*) and couch grass (*Triticum repens* or *Elymus repens*) are indicative of a fertile soil which is not compacted, allowing the rhizomes to spread. Of the two, stinging nettles favour a damper soil. Deep-rooted weeds such as the dandelion (*Taraxacum officinale*) and the dock (*Rumex obtusifolius*) favour a heavy poorly-drained soil. Cow parsley (wild chervil, *Anthriscus sylvestris*) tends to grow in a neutral soil that remains moist but does not become completely water-logged. The ox-eye daisy (*Chrysanthemum leucanthemum*) and the creeping buttercup (*Ranunculus repens*) can be found on neglected poorly-drained soils. Plantains (*Plantago sp.*) and mosses are found in poorly-drained lawns. The root systems of plantain and other weeds, such as creeping buttercup and daisies, which spread close to the surface, may be indicative of a compacted soil. The stolons will spread even when the soil has been trampled. Chickweed (*Stellaria media*) is a shallow-rooted weed which spreads rapidly across a reasonably fertile soil.

Weeds that have nitrogen-fixing bacteria associated with them, such as clover (*Trifolium repens*), bird's foot trefoil (*Lotus corniculatus*) and the vetches (*Vicia sp.*), are successful invaders on soil that is low in nitrogen. Various lupins can flourish as garden escapes on poor soil because they also have symbiotic bacteria associated with them that fix nitrogen. The periwinkles (*Vinca major* or *minor*) and ground ivy (*Glechoma hederacea*) spread rapidly as ground cover plants in damp, shaded conditions.

Mustard (*Brassica juncea*), poppies (*Papaver rhoeas*) and cornflowers (*Centaurea cyanus*) are widely distributed on chalk downland and are indicative of a shallow alkaline soil, whilst the perennial thistles (*Cirsium arvense* or *Sonchus arvensis*) and bladder campion (*Silene vulgaris*) require a deeper, alkaline soil. The mosses, sorrels (*e.g. Rumex acetosella*), the brackens and ferns and escaped rhododendron bushes require an acidic soil. Finally, a good look in the neighbourhood can be highly informative on the type of the soil.

When clearing weeds it is important to consider how they spread. Annual weeds need to be cleared before they set seed. Some may produce more than one crop in a season. If they are composted, the temperature of the compost heap needs to be above that required to kill the seeds otherwise this just becomes a mechanism for distributing the seeds around the garden in a fertile environment. Perennial weeds, such as couch grass and the bindweeds, spread through an active root system which needs to be carefully dug out. The smallest fragment of their conspicuous white root which is left behind, can initiate a new plant. Using a fork rather than a spade to clear the soil is clearly a more sensible approach. The bindweeds can be particularly deep-rooted.

2.12 HERBICIDES

Herbicides fall into a number of groups depending on their mode of action. Some are systemic and are translocated through the plant, others are hormone mimics and the final group is the contact herbicides. Whilst some are selective and act against broad-leaf weeds, others are non-selective and will affect any plant. Their persistence can be a matter of concern, particularly if the plant material is composted.

Glyphosate (Roundup® or Tumbleweed®; *N*-phosphonomethylgly-
cine) **2.13** is a systemic herbicide which is particularly useful for treating
deep-rooted perennial weeds. It blocks a key enzyme in the shikimic acid
pathway which catalyzes the reaction of shikimic acid 3-phosphate and
phosphoenolpyruvate to form 5-enolpyruvylshikimate-3-phosphate (see
section 1.9). This intermediate leads *via* chorismic acid to the aromatic
amino acids, phenylalanine and tyrosine. There is a similarity between
the structure of glyphosate and phosphoenolpyruvate. Glyphosate is at
its most effective in late spring, as the plant approaches flowering when
metabolic activity is at its greatest. It is absorbed by the leaves and on a
small scale the most effective treatment is to paint a solution onto the
leaves. Glyphosate is deactivated by the soil.

Glufosinate® (DL-phosphinothricin) **2.14** is a non-selective contact
herbicide which disrupts the formation of glutamine from glutamic acid
and ammonia, which is an essential step in the incorporation of nitrogen
into plant metabolites. There is a structural similarity between glutamic
acid and glufosinate. The peptide, phosphinothricylalanylalanine,
was originally isolated as an antibiotic from a *Streptomyces* species.
Acetic acid or the more wax-soluble, pelargonic acid (n-nonanoic acid),
are found in some contact herbicides. Formulations of the latter with
maleic hydrazide, succinic acid or lactic acid have been reported to be
safe non-selective herbicides. They appear to act by disrupting cell
membranes.

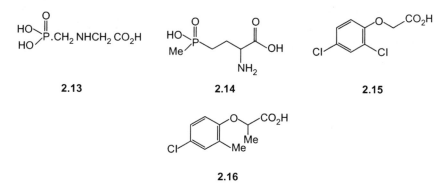

2.13   **2.14**   **2.15**

2.16

A large family of herbicides consists of mimics of the hormone,
indolylacetic acid. Many of these are selective for broad-leaf weeds. The
best known of these are 2,4-dichlorophenoxyacetic acid (2,4-D) **2.15** and
2-methyl-4-chlorophenoxyacetic acid (MCPA). Mecoprop® (MCPP) is
2-(2-methyl-4-chlorophenoxy)propionic acid **2.16** whilst dichloroprop®
is the 2-(2,4-dichlorophenoxy)propionic acid analogue. In these
chiral cases it is the *R* isomer which is active. The homologue,

(2,4-dichlorophenoxy)butyric acid is more selective and is only active against plants which have the ability to degrade the side chain to the phenoxyacetic acid. 2,4,5-Trichlorophenoxyacetic acid (2,4,5-T) achieved notoriety as a defoliant in 'Agent Orange' because of the presence of 2,3,7,8-tetrachlorodibenzo-*p*-dioxin as a contaminant. This tetrachloro compound is very resistant to biodegradation. Dicamba® is 3,6-dichloro-2-methoxybenzoic acid and is found in some preparations but because of potential toxicity problems it is not recommended for use in crop areas. Some pyridine analogues of the phenoxyacetic acids, such as triclopyr® and fluroxypyr®, are also used. However, the pyridine acids, clopyralid® **2.17** and aminopyralid®, persist in composted plant material and have caused problems in some commercial composts.

Another family of herbicides are dipyridinium salts exemplified by paraquat (*N,N'*-dimethyl-4,4'-bipyridinium dichloride) **2.18** and diquat (1,1'-ethylene-2,2'-bipyridylium dibromide). They act by inhibiting photosynthesis by accepting electrons from photosystem 1 and transferring them to oxygen to create destructive reactive oxygen species. There are toxicity hazards associated with the ingestion of the concentrated material. Sodium chlorate was used as a non-selective weed killer with a soil sterilizing effect but it has significant human toxicity and is no longer available. It was sometimes used in conjunction with atrazine (2-chloro-4-(ethylamine)-6-(isopropylamine)-*s*-triazine) **2.19**. The latter has been banned in the EU because of its persistent groundwater contamination. This persistence is slightly surprising because of a reactive imino chloride in the structure. This reactive group may account for its biological activity in binding to the proteins involved in the electron transport system in plants.

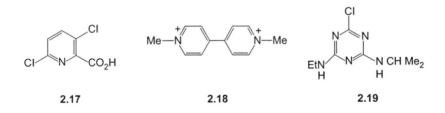

2.17   2.18   2.19

The toxicity of some herbicides has stimulated the search for novel lead compounds, particularly from amongst the allelopathic metabolites of plants. For example, the trione, leptospermone **2.20**, was identified as an allelopathic product of the bottle-brush plant, *Callistemon citrinus*. As a result, mesotrione **2.21** was synthesized which inhibits 4-hydroxyphenylpyruvate oxygenase. Volatile monoterpenes released by *Salvia* species, particularly 1,4- and 1,8-cineole, are potent allelochemicals.

Cinmethylin® **2.22**, an ether of 2-*exo*-hydroxy-1,4-cineole, has been developed based on these as a possible pre-emergence herbicide. Another compound to attract interest in this context is the pyrrolidine amide of (2*E*,3*E*)-decadienoic acid, sarmentine **2.23**, which has been isolated from a pepper, *Piper longum*.

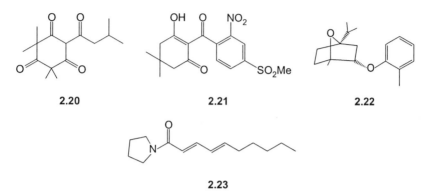

2.20 **2.21** **2.22**

2.23

2.13 GERMINATION OF SEEDS

A number of physicochemical features affect the successful germination of vegetable seeds. During the germination of a seed, endospermic reserves surrounding the embryo are gradually mobilized to facilitate the emergence of the seedling. Since seeds have a low water content to reduce the likelihood of premature germination, this mobilization requires the absorption of water. Some seeds are covered with a thick testa, or a wax coating, and hence an overnight soaking, *e.g.* of beans, may facilitate germination. The absorption of water facilitates the migration of a gibberellin plant hormone, such as gibberellic acid or gibberellin A_1, to the metabolically active aleurone layer of the seed where it induces α-amylase to release glucose from starch. Very small amounts of gibberellic acid are required to speed up the germination of vegetable seeds.

The successful germination of seeds is both temperature and light dependent. The majority of seeds are produced in the autumn and they require a cold period in their dormancy before germination in the spring. This cold period can be mimicked by storing the seed in the refrigerator. The requirement can also be overcome by treatment with gibberellic acid. Optimum germination is normally in the temperature range 7–16 °C and hence for sowing in the soil, it is important not to be too hasty in the spring. The storage of seed is very important in obtaining a successful rate of germination. Hardware shops that sell seeds sometimes display their colourful packets near the window where they are subjected to quite high temperatures in the sunlight. This can reduce the viability of these seeds.

Germination is light dependent and is promoted by red light of a wavelength of approximately 660 nm and inhibited at 720 nm. Chlorophyll in the developing seedlings absorbs light at around 660 nm. Consequently, when the leaves from other plants shade an area, germination below them is not promoted. Germination may also be restricted by intense sunlight.

A number of allelopathic compounds inhibit germination. The naphthoquinone, juglone **2.24**, which is produced by the walnut tree and monoterpenes, such as menthone and cineole, produced by mints and rosemary are examples. Some seeds, such as carrots, produce their own germination inhibitors such as crotonic acid, restricting the density of germinated seeds and thus the competition for nutrients. Other compounds can stimulate germination. There is sometimes a surprisingly rapid regeneration of plant life after a heath fire, or when the stubble is burnt in fields in late August. Smoke from burning vegetation is used to promote germination of seeds, for example of heather, in restoring moorland. Bubbling smoke through water is a convenient way of producing this effect. The active compound has been identified as a butenolide, karrikinolide **2.25**, and it has been shown to be effective with a number of vegetable crops such as lettuce. However, high concentrations of smoke in the water are inhibitory, an effect attributed to a second butenolide. The presence of the butenolide ring in these compounds bears a similarity to that of strigol **2.26** which stimulates the germination of the parasitic *Striga* weeds.

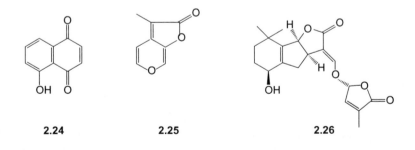

2.24   2.25   2.26

2.14   INSECT PESTS

The control of insect pests has had an immense impact on food production. However, insects have both detrimental and beneficial effects on plants. The development of a sustainable strategy in the kitchen garden to maintain a balance between these effects is a topic of continual study. Whilst herbivorous insects not only destroy plants and spread microbial disease, other insects are pollinators and some have a beneficial effect on soil structure and facilitate the integration of compost into the soil. Furthermore, insects play an important role in the natural food chain for birds and mammals. Over the last few years, ecological and health issues have led to the withdrawal of a significant number of insecticides, particularly from the domestic market, including the organochlorine insecticides such as DDT **2.27**, aldrin **2.28**, dieldrin and lindane. The chlorine atoms, whilst required for their biological activity such as the effect on ion channels in insect nerve cells, also blocked metabolism and increased lipid solubility, leading to persistence in the environment and in the food chain. Nevertheless the role of, for example, DDT in the control of insect vectors of disease should not be underestimated. Various organophosphates and carbamates, which bind to acetylcholine esterase and interfere with insect nerve impulses, have human toxicity problems associated with them. Some of these have also disappeared from the domestic horticultural market.

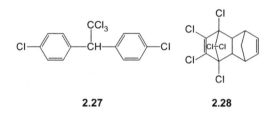

2.27   **2.28**

An understanding of specific insect life cycles, for example where and in what form they over-winter and what triggers their predatory response to plant development, can provide useful control measures within the kitchen garden. Appropriate winter washes, tidiness, greenhouse fumigants and physical barriers have an important part to play. There are various chemically mediated stages in the insect life cycle that can form the focus for insect control, whether as feeding deterrents, metabolic inhibitors or in the disruption of metamorphosis. There are also volatile chemicals produced by the damaged plant or by the pest that attract predators of the insect.

Plants have evolved their own means of controlling insect attack. Two families of insecticides based on natural products are the pyrethroids and, more recently, the neonicotinoids. The use of pyrethrum (*Tanacetum* or *Chrysanthemum cinerariifolium*) as an insecticide has been known for centuries. The identification of the active constituents as monoterpenoid esters (*e.g.* pyrethrin 1 **2.29**) led to a widely used family of insecticides exemplified by permethrin **2.30** or deltamethrin. These are sometimes used with piperonyl butoxide **2.31** as a synergist.

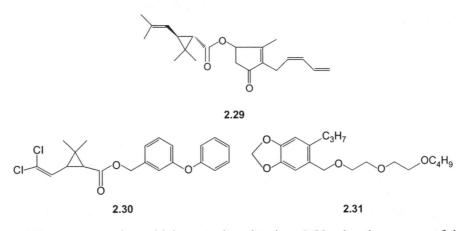

2.29

2.30   **2.31**

Nicotiana species which contain nicotine **2.32** also have a useful insecticidal effect. Liquid tobacco extracts were used in the eighteenth and nineteenth centuries to control pests found on fruit trees and in the greenhouse. A soap solution was sometimes added to these extracts to increase their penetrating power. Nicotine itself has a well-known mammalian toxicity. A series of neonicotinoids based on this structure have been developed. Nevertheless one of the best known of these, imidacloprid **2.33,** is under suspicion because of its possible impact on the honey-bee population.

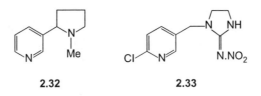

2.32   **2.33**

Another biological source of an insecticide is derris root (*Derris elliptica*) in which the active components include rotenone, sumatrol and elliptone. Rotenone interrupts electron transport in the mitochondria and there are now restrictions on the use of derris powder.

Marigolds (*Tagetes* species) have a marked insecticidal activity associated with the formation of phenylhepatriene and a series of thiophenes. α-Terthienyl **2.34** is a root-exudate which is particularly active against root-knot nematodes. There are reports of a reduction in root lesions arising from nematode attack on potatoes when the potato crop is preceded by marigolds. The marigold has been recommended as an 'inter-cropping' plant. The insect deterrent activity of the organosulfur compounds produced by *Allium* species such as chives makes them a possible 'inter-cropping' plant.

A number of monoterpenes, such as limonene, citronellol and thymol, have useful insecticidal effects and herbs which produce these, or extracts from them, have potential uses. It is worth remembering that the monoterpenes which are produced by citrus fruits that grow in warmer Mediterranean climates with a higher insect population have this role. Extracts of the peel of these fruits and of the leaves of plants from the Lamiaceae family have potential insect deterrent activity. Neem extract from *Azidirachta indica*, which contains azidirachtin, is of interest in this connection as well.

Slug control by the molluscicide, metaldehyde, is a common line of defense against the destructive ability of these pests. Numerous other methods have been recommended for their control, including the use of residues from 'spent' coffee grounds which retain caffeine. A carbamate, methiocarb **2.35**, derived from 4-methylthio-3,5-dimethylphenol, is quite active as a molluscicide.

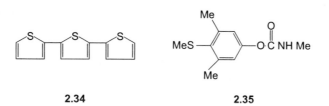

2.34   **2.35**

Within the limited space of the kitchen garden, a strategy in planting can reduce insect damage. Many insects are attracted to their host plant by specific volatile organic compounds which the plant produces particularly if it has been damaged. In the wild, plants do not normally grow in blocks in neat rows all emitting volatile attractants at the same time and en masse. Mixed planting, inter-cropping and spreading the time of sowing and a careful consideration of neighbouring shrubs, can all have an underlying chemical justification.

CHAPTER 3

The Chemistry of Root and Stem Vegetables

The roots, tubers and bulbs which are used as vegetables, are the storage organs of plants that may be biennials or perennials. During the first year of growth, these parts of the plant are enlarged to contain carbohydrates in the form of sugars and starches in order to provide energy reserves. As roots, tubers and bulbs have to provide for the start of a new season's growth, they also contain proteins, vitamins and plant growth hormones, together with defensive natural products to afford protection against predators. In this chapter, we will consider vegetables of this type that are commonly grown in the UK. These include potatoes, carrots, parsnips, beetroots, onions, garlic, leeks and some other stem vegetables, such as asparagus, celery and rhubarb. Many of these vegetables belong to different plant families and consequently they contain a variety of different natural products.

3.1 POTATOES

The potato, *Solanum tuberosum*, is a common garden and allotment crop and often forms part of a regular crop rotation. The potato belongs to the Solanaceae which is the same family as the tomato and so, not unexpectedly, there is some overlap in their constituents. It is also the same family as the deadly nightshade. The potato was brought to Europe from South America in 1536. Its origins have been traced back to Chile and Peru where it had been cultivated for many thousands of years. A number of related species are found growing in this region.

Chemistry in the Kitchen Garden
By James R. Hanson
© James R. Hanson 2011
Published by the Royal Society of Chemistry, www.rsc.org

Examination of varieties of *S. tuberosum* growing in the Andes and of other related species has revealed examples which may have higher levels of beneficial natural products. The potato is now grown throughout the world with substantial crops being produced not just in Europe but also in India and China. Indeed some potatoes sold in UK supermarkets are produced in Egypt from Scottish seed potatoes. In this country, we each consume about 30–40 kg of potatoes per annum.

There are many varieties of potato which have been bred for different characteristics of tuber formation including time of production (early or maincrop), disease resistance, flavour and consistency. Although we sow what is described as a 'seed' potato, it is not a genuine seed. The latter is produced in a small poisonous fruit which sets after flowering. The potato plant is a perennial which dies back each year and the potato tuber represents the storage organ, hence its high carbohydrate content. Reproduction of the plant in the garden or on the allotment is vegetative.

A typical potato may weigh between 100–150 g, although salad and early potatoes are smaller. It contains about 75% water. The current popularity for growing potatoes in containers on the patio can reveal their significant requirement for water. Whilst much of the bulk of the potato is carbohydrate, there is a protein layer under the skin and in the sprouts. A typical raw potato contains about 20 g carbohydrate of which 15 g is starch. Starch is a generic term covering a range of polysaccharides of glucose with different molecular weights. Starch comprises two major components, amylose and amylopectin. Amylose is a linear [1α-4]-polymer of glucose which can take up a helical structure, whilst amylopectin has the same backbone but contains chain branching units attached to C-6 (the primary alcohol) of glucose. A typical ratio of amylose to amylopectin is 1 : 4, although this can vary. A potato with a high amylopectin content is described as being 'waxy'. A much higher amylopectin content is of value in commercial applications such as paper-making and adhesives. A genetically modified potato, Amflora®, in which the amylopectin content is almost 100%, has received EU approval to be grown for non-foodstuff uses.

Although much of the starch is degraded to glucose within the stomach thus accounting for the high glycemic index of the potato, a portion, which can be as much as 13% after cooking, is resistant to this degradation and enters the large intestine. The proportion of resistant starch can vary with the cultivar and it can function as dietary fibre. Apart from acting as a carbohydrate and hence as an energy source, the potato is a source of a number of vitamins, such as vitamin C (20 mg per 100 g) and small amounts of vitamins B_1, B_2, B_3 and B_6, together with some protein and traces of metal ions.

The potato tubers contain a number of polyhydroxyphenols of which chlorogenic acid **3.1** is the most important, representing as much as

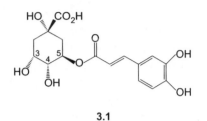

3.1

90% of these compounds. Chlorogenic acid is the C-5 caffeic acid (3,4-dihydroxycinnamic acid) ester of quinic acid. Other potato polyphenols are different esters of quinic acid. Thus neochlorogenic acid and cryptochlorogenic acid are respectively the C-3 and C-4 caffeic acid esters of quinic acid. The amounts of these polyphenols that are present vary throughout the plant. On average the potato sprouts contain about 750 mg per 100 g fresh weight and the leaves contain over 200 mg per 100 g, whilst the roots contain about 25 mg per 100 g and the tubers about 17 mg per 100 g where it is located just below the skin. These compounds serve as protective agents. The polyphenol oxidase converts the dihydroxyphenol to an orthoquinone which will react with nucleophiles or polymerize. These oxidation products are toxic to pathogens and their polymers seal wounds. There is some evidence that the polyhydroxyphenols accumulate in sound tissue adjacent to injured tissue and their polymeric products may then form a wound barrier layer. A corky scab which forms around a wound contains suberin, which consists of these aromatic polymeric products together with aliphatic polymers based on unsaturated fatty acids. In the leaf, the polyhydroxyphenol oxidation products are associated with resistance to insect attack. Chlorogenic acid can have health benefits as an antioxidant. It is worth noting that these compounds are very water soluble and hence they can be lost in boiling water. Cooking new potatoes in their skins is a way of retaining some of their beneficial properties.

The oxidation of the polyphenols by polyphenol oxidase is associated with the enzymatic browning of potatoes. The internal discolouration of potatoes induced by bruising has been linked to the oxidation of the phenolic amino acid, tyrosine **3.2**, *via* an orthoquinone and the formation of melanin polymers. Traces of iron(II) in the potato can also lead to a grey discolouration of potatoes when they are cooked. The iron(II) forms a complex with chlorogenic acid which is oxidized to the dark iron(III) complex.

The potato contains a group of steroidal glycoalkaloids. The two major glycoalkaloids are α-chaconine and α-solanine, which are both glycosylated derivatives of the aglycone, solanidine **3.3**. Like the chlorogenic acids, the glycoalkaloids are unevenly distributed in the plant. Typical concentrations are (per 100g fresh weight): tubers (2–15 mg), stems (30–45 mg), roots (85 mg), leaves (145 mg) and sprouts (997 mg). Quite high concentrations (approx. 45 mg) are also found in the berries and seeds. The alkaloid concentration is highest in those parts of the plant that are the most vulnerable to insect attack. The alkaloids can accumulate in potatoes where the skin has turned green and is exposed to light. Hence when storing potatoes, it is important to ensure that they are not exposed to light. Any 'greened' potatoes should be rejected. The alkaloids are toxic to man and have a haemolytic effect and some may be teratogenic. Solanine also has choline esterase inhibitory activity. The alkaloids possess antimicrobial activity and there is some evidence that their formation can be stimulated by microbial attack. The sterols which have been identified in the leaves of the potato include those that are involved in the biosynthetic sequence from cycloartenol to cholesterol, and are precursors of the alkaloids.

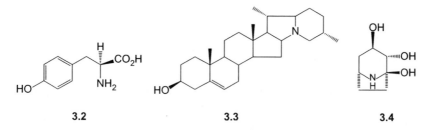

3.2   3.3   3.4

Two water-soluble nortropane alkaloids, calystegine A_3 **3.4** and B_2 which are powerful glycosidase inhibitors, have also been found in the peel of potatoes. These alkaloids are related to the tropane alkaloids, such as atropine and scopolamine, which have been found in other members of the Solanaceae, *e.g. Datura* species.

The formation of the tubers is controlled by photoperiod and it is stimulated by the production of tuberonic acid **3.5** in the leaves. This compound is a relative of the plant hormone, jasmonic acid **3.6**, which will also stimulate tuber formation. Potato tubers do not sprout immediately after harvest but require a period of winter dormancy before they are 'chitted' and buds appear. This rest period can be shortened by treating the freshly harvested potatoes with a very dilute solution (1 mg l^{-1}) of the plant hormone, gibberellic acid. The natural gibberellins which have been found in potatoes include gibberellins A_1 **3.7**, A_{15} **3.9**, A_{19} **3.10** and A_{20} **3.8**. Gibberellin A_{15} occurs as the $19 \rightarrow 20$-lactone rather than as a hydroxy acid. They have been detected in vigorous sprout growth in concentrations of 0.5–0.6 ng per g fresh weight.

The sprouting of potatoes in storage is inhibited by compounds such as chloropham® (isopropyl-*m*-chlorophenylcarbamate). However, it can also be reduced by using volatile essential oils from mints such as spearmint or peppermint oil soaked on blotting paper. Even mint leaves have been recommended. It has been suggested that the monoterpene, (*R*)-carvone, is the active compound. The monoterpenes produced by mints are well-known as germination inhibitors.

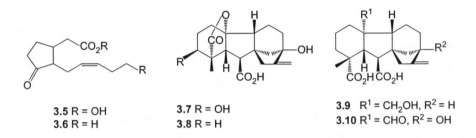

3.5 R = OH
3.6 R = H

3.7 R = OH
3.8 R = H

3.9 R^1 = CH$_2$OH, R^2 = H
3.10 R^1 = CHO, R^2 = OH

Typical of higher plants, the leaves of the potato are covered with a cuticular wax which has a primary role of reducing water loss and also of affecting the resistance of the plant to herbivores. The principal components of the waxes are long chain (C$_{23}$–C$_{34}$) n-alkanes in which the C$_{31}$ hydrocarbon predominates. The typical concentrations are 3–5 μg.cm^{-2}. Some fatty acids and their esters were also detected. A number of volatile sesquiterpenes are also released by potato leaves. These include the hydrocarbons, humulene **3.11**, caryophyllene **3.12**, germacrene D **3.13**, and the alcohols, germacrene D 4-ol **3.14** and α-cadinol **3.15**. As many as 250 compounds have been identified in the volatiles produced by potato tubers during cooking. These include the methyl esters of 3-methylbutanoic, 3-methylbutenoic and 2-methylpropanoic acids, 3-methylbut-2-en-1-ol and butylfuran, together with the sesquiterpene, α-copaene. Methional [3-(methylthio)-propanal] which is also formed from methionine, is also present in small amounts. However, it can be smelt at the part per billion level and it may account for some of the unpleasant smell of decomposing potatoes. 3-Ethyl-3,6-dimethylpyrazine has also been identified.

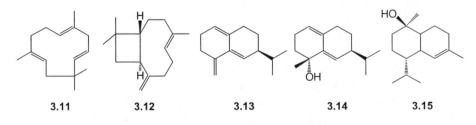

3.11   3.12   3.13   3.14   3.15

Potatoes are attacked by a number of insects, fungal and bacterial pathogens. One of the most serious insect pathogens, although not yet present in the UK, is the Colorado beetle (*Leptinotarsa decemlineata*). Unlike other insects, it is not deterred by the glycoalkaloid, solanine. A plant which is related to the potato, *Solanum demissum*, is resistant to attack. This has been associated with the presence of the steroidal alkaloid, demissine. There are subtle structural differences between this alkaloid and solanine, including the presence of four sugars rather than three and the reduction of the 5,6-double bond. Attempts have been made to breed potatoes containing this alkaloid. The beetle is attracted to the potato by the volatile sesquiterpenes, particularly caryophyllene **3.12**. The aggregation pheromone which is produced by the male is the highly oxygenated monoterpene, (*S*)-1,3-dihydroxy-3,7-dimethyl-6-octen-2-one **3.16**. Bioassays of synthetic material have revealed that the (*S*)-isomer is attractive to both the male and female beetles while the enantiomer is inactive. The chiral aggregation pheromone has been synthesized from 2,3-epoxynerol and has been used to bait traps for the Colorado beetle. It is possible that the beetle produces this compound by modifying (*S*)-linalool obtained from the plant. The widespread use of insecticides in the USA has led to resistance developing in the Colorado beetle although it is susceptible to microbial insecticides based on *Bacillus thuringiensis* (Bt). A number of insect antifeedants from other plants, such as some sesquiterpenoid silphinene derivatives from *Senecio palmensis*, have also been examined in this context. A stinkbug, *Perillus bioculatus*, is a predator of the Colorado beetle and is attracted to damaged plants by the emission of humulene **3.11**, caryophyllene **3.12**, (*E*)-β-farnesene and germacrene D **3.13** and its 4-alcohol **3.14**. The maggots of a tachinid fly, *Myiopharus doryphorae*, are known to parasitize the larvae of the Colorado beetle and may also be useful as a biocontrol agent. Aphids can also be a problem on potatoes and are deterred by the chlorogenic acid **3.1** and glycoalkaloid content of the leaves and stems.

Another cause of economic loss of potatoes is the potato root eelwom, *Heterodera* (*Globodera*) *rostochiensis* sometimes known as *Globodera pallida*. The potato produces a highly oxidized triterpene, solanoeclepin A **3.17**, which diffuses into the soil and at nanomolar concentrations stimulates the hatching of the potato cyst nematodes. The development of these nematodes in the soil is a reason for ensuring crop rotation. A genetically modified potato has been developed to combat the potato cyst nematode. The Desirée potato cultivar has been modified to produce a particular peptide cystatin, which is a cysteine protease inhibitor. This peptide has 115 amino acid residues and the gene for its production was derived from rice. By inhibiting the cysteine proteases, it renders the potato skin less digestible by the nematode reducing the severity of the attack.

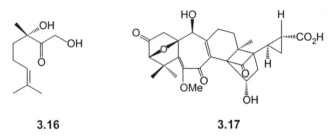

3.16   3.17

Further economic losses are caused by the action of fungal and bacterial pathogens. The most serious of these is the potato late blight, *Phytophthora infestans*, a virulent strain of which brought about the Irish potato famine of 1845. The spores of the fungus are spread by the wind and may also remain in the soil. The fungus spreads in warm wet weather particularly from late May onwards, a feature which provides a good reason for growing early potatoes. The infection first appears as brown spots on the leaves which spread rapidly. Chlorophyll is lost and hence sugar and starch production stops and the tubers cease to grow. In a bad year they may also then suffer from a secondary bacterial infection. When the potato is attacked by the fungus, it produces the sesquiterpenoid phytoalexins, rishitin **3.18** and lubimin **3.19**. Virulent strains of fungal pathogens including amongst others, such as *Gibberella pulicaria*, can detoxify these antifungal agents by epoxidizing the alkenes. The antifungal coumarin glycoside, scopolin **3.20**, is also produced around the wounds. Some *Solanum* species such as *S. venturii* and *S. mochiquense* are resistant to attack by *Phytophthora infestans*. Genetically modified red-skinned Desirée cultivars containing the resistant genes have been grown in field trials. The Desirée potato is used because the red skin provides a means of identifying the modified plants. The transgenic potato recognises specific molecules produced by the pathogen and this triggers the natural plant defences.

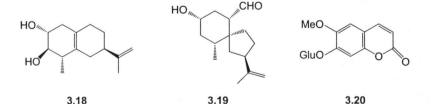

3.18   3.19   3.20

The soft rot bacterium *Erwinia carotovora* (*Pectobacterium carotovorum*) can cause a serious post-harvest loss of potatoes, whilst the common scab of potatoes is produced by another soil bacterium, *Streptomyces scabies*. In response to this attack, the potato produces a protective layer of phenolic compounds and fatty acids known as suberin, at the site of infection. *Streptomyces scabies* produces a group of phytotoxins which are known as the thaxtomins (*e.g.* thaxtomin A **3.21**). These diketopiperazines are derived from L-4-nitrotryptophan and L-phenylalanine. The presence of a nitro group in a microbial metabolite is quite unusual. These compounds are phytotoxic at nanomolar concentrations and cause the collapse of cells in the potato tuber. This bacterium can also elicit scab symptoms on other root crops such as carrots, parsnips, turnips and beetroots. The persistence of this bacterium in the soil is a point to remember when planning crop rotations.

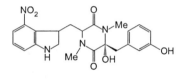

3.21

Examination of varieties of *Solanum tuberosum* growing in the Andes has revealed examples which produce the carotenoids, lutein and zeaxanthin, as well as β-carotene **3.22**. These have led to a number of pigmented cultivars. The antioxidants α-tocopherol and chlorogenic acid were found in quite high levels in some of the varieties. A dark purple fleshed variety contained the anthocyanin petanin (petunidin **3.23** 3-coumaroylrutinoside-5-glucoside).

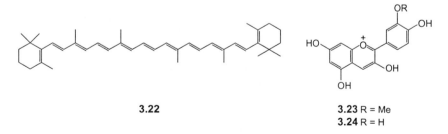

3.22

3.23 R = Me
3.24 R = H

Other edible tuberous roots which include the name 'potato' such as the sweet potato (*Ipomoea batatis*) and the Livingstone potato (*Plectranthus esculentus*) are quite different species and have a different chemistry.

3.2   CARROTS

The carrot (*Daucus carota* subsp. *sativa*) is a popular root vegetable. The plant is a member of the Apiaceae (Umbelliferae) and it is supposedly a cultivated form of the wild carrot, (*Daucus carota* subsp. *carota*). Although the carrot is a biennial, the edible part is the enlarged taproot which is harvested during the first year when it is fulfilling its storage function and before extensive lignification has taken place which generates a woody carrot. The ancestors of the carrot came from Afghanistan and there is evidence of it being cultivated in the Ancient World before reaching Europe in the Middle Ages. Early forms of the carrot were purple or deep red. Plant breeders in the Netherlands in the 16th and 17th century are accredited with developing the present-day orange carrot.

Carrots are best known for the orange carotenoid pigments which were first isolated from carrots in 1831 by Wackenroder. The chromatographic separation by Kuhn of carotene in 1931 into its isomers of which β-carotene **3.22** was the major component, provided a significant demonstration of the power of chromatography. The total carotenoid content of the carrot depends on the cultivar but is about 40 mg per 100 g dry weight, of which about 80% is β-carotene. A yellow complex of β-carotene and a protein has been detected in carrots. Carotenoid–protein complexes are well-known in lobsters where their disruption on cooking leads to the development of the orange-red pigment of lobster. The amount of available carotenoid rises as carrots are cooked.

The carotenoids have a well-documented range of beneficial activities as antioxidants and free-radical scavengers. Their consumption has been correlated with a reduced risk of developing certain cancers although excess can be detrimental to the liver. β-Carotene has attracted particular interest because of its pro-vitamin A activity since it is converted into retinal **3.25** by mammals. There was a war-time story that eating large quantities of carrots enhanced the night vision of RAF crew in combating enemy planes, a legend which popularised the vegetable. Apart from increased concentrations of carotenoids, the colouring matters of the purple carrots include an anthocyanin, a glycoside of cyanidin **3.24**, which may also confer further beneficial properties on the carrot.

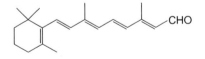

3.25

The major components of the volatile oil obtained from carrot roots were the monoterpenes, myrcene **3.26**, terpinolene **3.27**, sabinene **3.28**, and the aromatic hydrocarbon, *p*-cymene **3.29**, together with the sesquiterpenes, γ-bisabolene **3.30**, humulene **3.11** and caryophyllene **3.12**. The presence of the monoterpenes serves as attractants for the carrot root fly. The aldehydes, 2-nonenal and octanal, also make a contribution to the taste. Carrot seed oil contains some different sesquiterpenoid constituents including carotol **3.31** and daucol **3.32**. The sweet taste of carrots comes not just from the presence of sugar but also from β-ionone **3.33** which is a degradation product of β-carotene. Sucrose followed by glucose and fructose are the main sugars that are present.

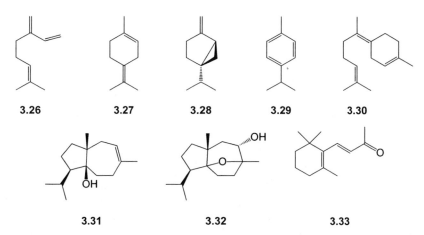

Crotonic acid [(*E*)-but-2-enoic acid] has been identified in aqueous extracts from carrot seed. This acid has an allelopathic effect and inhibits the germination and development of young seedlings. It has been shown that too dense a sowing of carrot seed results in delay and inhibition of their germination, an autotoxic effect which has been attributed to crotonic acid. This provides a natural means of controlling plant numbers presumably in order not to exhaust nutrients.

When carrots are challenged under various stress conditions such as attack by insect and microbial pathogens, they synthesize phytoalexins, such as scopoletin **3.34**, eugenin **3.35** and 6-hydroxymellein **3.36**, as well as some antifungal polyacetylenes, such as falcarinol **3.37** and falcarindiol **3.38**. These compounds, particularly falcarindiol, are responsible for the bitter taste of infected carrots. A number of phenolic compounds, such as chlorogenic acid and other cinnamic acid derivatives, together with lignin and suberin are also formed by carrots as wound barriers.

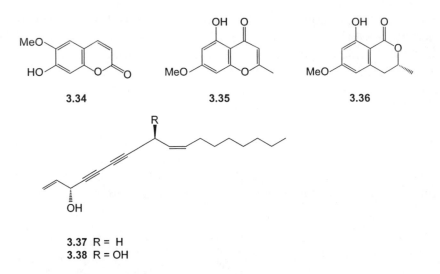

3.34   **3.35**   **3.36**

3.37 R = H
3.38 R = OH

Despite their bitter taste, the polyacetylenes, falcarinol and falcar-indiol, have some beneficial effects. Not only do they show antibacterial activity but falcarinol also shows cytotoxic activity against some human cancer cell lines and they may have a chemo-protective effect against various tumours. These compounds are also found in other vegetables from the Apiaceae, such as the parsnip and roots of celery. Falcarinol may be present to the extent of 25 mg per kg fresh weight. However, the extent to which these compounds leach out during cooking can reduce their effectiveness. Indeed it has been suggested that carrots should be cooked whole and cut up afterwards to retain their beneficial compo-nents. These C_{17} polyacetylenes are biosynthesized from the C_{18} acid, crepenynic acid [$CH_3(CH_2)_4C{\equiv}CCH_2CH{=}CH(CH_2)_7CO_2H$], by desa-turation, decarboxylation and hydroxylation.

3.3   PARSNIP

The parsnip (*Pastinaca sativa*) is also a member of the Apiaceae. Like the carrot, it was cultivated in the Ancient World and there are refer-ences to the medicinal properties of the plant in the writings of Pliny and subsequently in the herbals of Culpepper and Gerard in the Middle Ages. The parsnip is a biennial and the taproot is harvested during the winter. Indeed the flavour is allegedly enhanced by exposure to frost and the consequent rupture of some of the plant cells.

Parsnips contain starch as well as the free sugars, glucose, galactose, fructose and sucrose. When infected by fungi or damaged by insects they produce a series of furanocoumarin phytoalexins, such as angelicin **3.41**, psoralen **3.39**, xanthotoxin **3.40** and sphondin **3.42**. The extra carbon

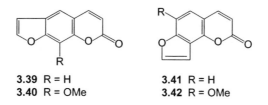

3.39 R = H
3.40 R = OMe

3.41 R = H
3.42 R = OMe

atoms of the furanocoumarins arise from an isoprene unit which is attached to the coumarin and then degraded. These furanocoumarins are phototoxic. When they are absorbed by the skin and activated by sunlight, they produce contact dermatitis. There have been a number of reports of contact dermatitis in agricultural workers which have been traced to handling infected parsnips. Some weeds, such as cow parsley and the giant hogweed which also belong to the Apiaceae, can also cause this problem. These compounds are toxic and behave as antifeedants to a range of polyphagous insects. The parsnip webworm, *Depressaria pastinacella* which routinely consumes plant material containing these compounds, has evolved an interesting mechanism to cope. It sequesters the carotenoid lutein **3.43** from the plant. The light-absorbing properties of this compound give the insect protection against the photo-oxidative stress associated with ingestion of the photoactive furanocoumarins. The parsnip webworm then has time to metabolize the furanocoumarins by oxidatively cleaving the furan ring.

The major components of the essential oil of parsnip roots are myristicin **3.44** and the monoterpene hydrocarbon, terpinolene **3.27**. Myristicin, better known as the pro-psychotropic agent from nutmeg, has insecticidal properties. Parsnips also contain the cytotoxic polyacetylenes, falcarinol **3.37** and falcarindiol **3.38**, which were described in the previous section on carrots. It is possible that these compounds may exert some beneficial properties in the consumption of parsnips.

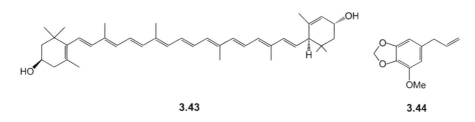

3.43

3.44

3.4 BEETROOT

The beetroot, *Beta vulgaris*, is a member of the Amaranthaceae family of the Caryophyllales. It is related to a number of *Chenopodium* (Goose-foot and Fat Hen) species. The cultivated beetroot was probably derived from the sea beet (*Beta vulgaris* subsp. *maritima*) and it is interesting in this connection that the major deep-red pigment is water soluble. The beetroot has a long history and was cultivated in the Mediterranean area in ancient times. Some varieties have been grown for the leaves (spinach beet or chard) whilst others have been grown for their tuberous root. Sugar beet is a non-pigmented version which has been grown commercially since the late 18th century as a source of sucrose. In modern cultivars the sucrose content can be as high as 15%. The garden beetroot contains 5–10% sucrose. The biosynthetic pathway leading to sucrose in sugar beet differs from that in sugar cane, a C_3 pathway in sugar beet *versus* a C_4 pathway in sugar cane. The carbon isotope (^{12}C *versus* ^{13}C) effects are sufficiently different in the two pathways to enable sugar and sugar products derived from sugar beet to be distinguished from those of cane sugar by measuring the carbon isotope ratios.

The tuberous root of beetroot which is grown as a garden crop forms a storage organ for the biennial plant. The major distinctive deep-red, water-soluble pigment is the glycoside, betanin **3.45**. This nitrogenous pigment belongs to a family of natural products known as the 'betalains', which comprise the red-violet betacyanins and the yellow betaxanthins. The underlying skeleton of these pigments contains an aldimine which is formed by the condensation of the aldehyde of betalamic acid with the nitrogen of a cyclo-dihydroxyphenylalanine (DOPA) or an amino acid such as proline. The former leads to the deep-red betanin **3.45** whilst the latter gives a betaxanthin, indicaxanthin **3.46**. Despite the intensity of the colour, betanin is present at a concentration of only about 0.5 mg per g fresh weight. There are albino varieties of beetroot which lack this pigment and some yellow-orange varieties which contain vulgaxanthin **3.47** in place of betanin. This is the iminium conjugate of betalamic acid with glutamine. Betalain pigments are widespread throughout the *Caryophyllales* and they have also been detected in some higher fungi such as the fly agaric, *Amanita muscaria*. Some beneficial antioxidant properties have been associated with the presence of these pigments. However, some people are unable to metabolize these pigments and they can cause discoloured urine.

Both portions of betanidin are derived from 3,4-dihydroxyphenylalanine (L-DOPA). Betalamic acid arises by cyclization of 4,5-secoDOPA whilst the cycloDOPA, with which it condenses, arises from dopaquinone.

An unusual non-protein amino acid, azetidine-2-carboxylic acid **3.48**, has been detected in beetroot although the amounts are quite low. This L-amino acid can be incorporated into some proteins in place of

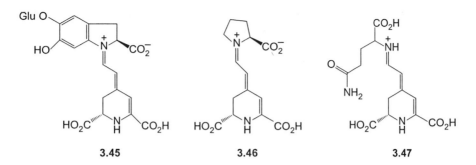

3.45　　　　　　**3.46**　　　　　　**3.47**

L-proline. Another characteristic of beetroot is the 'earthy' flavour which has been attributed to the presence of the norsesquiterpenoid, geosmin **3.49**. Although this is well known as a soil *Streptomycete* metabolite, there is evidence that in this case, it is biosynthesized by beetroot rather than being picked up from the soil.

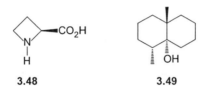

3.48　　　　　**3.49**

3.5　RADISH

The radish (*Raphanus sativus*) is a member of yet another plant family, the Brassicaceae family. It is a rapidly germinating seed and the derivation of the name 'Raphanus' (Gk. quickly appearing) indicates this. The red pigments are acylated anthocyanins based on the anthocyanidin, pelargonidin **3.50**. The pelargonidin glycosides, particularly the glucose units attached to C-3, are also acylated with various combinations of *p*-coumaric, ferulic and malonic acids. Typical of the Brassicas, radishes contain glucosinolates which release isothiocyanates by the action of myrosinase. Thus the major glucosinolate of radishes is glucoraphasatin **3.51** which after enzymatic hydrolysis by myrosinase, affords 4-methylthio-3-butenyl isothiocyanate (raphasatin, **3.52**). This compound which accounts for some of the strong taste of radishes, has powerful antioxidant activity and shows selective cytotoxic activity towards some human colon cancer cell lines. The dihydro derivative (4-methylthiobutylisothiocyanate) is erucin which is obtained from rocket (*Eruca sativa*) and the sulfoxide (4-methylsulfinylbutylisothiocyanate) is sulphorane which is obtained from broccoli (*Brassica oleracea*). The latter has attracted interest because of its anticancer activity and inhibitory effect on the growth of *Helicobacter pylori*. This bacterium is associated with the development of stomach ulcers.

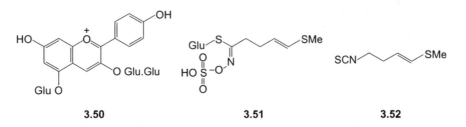

| 3.50 | 3.51 | 3.52 |

Turnips (*Brassica rapa* varn. *rapa*) and swedes are also bulbous roots plants that are members of the Brassicaceae. They contain glucosinolates which release isothiocyanates by the action of myrosinase.

3.6   BULB AND STEM VEGETABLES

The bulbs and stems of a number of plants have been used as vegetables. These bulbs act as storage organs for biennial plants and thus they must survive a winter period. Consequently they have evolved various protective measures including the biosynthesis of natural products which act as feeding deterrents for insects and small mammals as well as antimicrobial agents. Furthermore, the bulbs may be exposed to aerial and photo-oxidation and thus some of the constituents of the outer skin are light-absorbing and deactivate various reactive oxygen species. The Alliums including onions, shallots, garlic and leeks, fall into this class. Asparagus, celery and rhubarb are also plants that are grown for their stems and will be considered in this section.

The cultivation of the Alliums for use as vegetables has a long history. Onions and garlic may have originated in Central Asia. There is evidence for their cultivation in Egypt and there are references in the Old Testament of the Bible to their use as foodstuffs and to their beneficial properties. Throughout the Middle Ages there are records of the medicinal properties of these plants, as well as to the odours arising from their consumption. Many of these properties are associated with the chemistry of their sulfur-containing metabolites.

The organic chemistry of sulfur is quite complex. Although the majority of the natural products containing sulfur that are found in the Alliums arise from C_3S units derived from sulfur-containing amino acids, their structures are quite diverse. The reactivity of sulfur means that there is a diffuse borderline between a genuine natural product and an artefact arising from a chemical reaction after the substance has left the plant. It is therefore useful to review a few general points of sulfur chemistry. Sulfur is a poorer hydrogen bond acceptor than oxygen and consequently its compounds are often more lipophilic and volatile than their oxygen counterparts. On the other hand, the thiol S-H bond is more acidic than the hydroxyl group. Thiols (RSH) and sulfides (RSR) are powerful nucleophiles and readily form metal complexes. Sulfur exhibits several different formal valency states. Thiols (RSH) are readily oxidized to disulfides (RSSR) and to sulfenic acids (RSOH), whilst alkyl sulfides (RSR) give sulfoxides (R_2SO), sulfones (R_2SO_2) and thiosulfonates ($RSSO_2R$). A sulfoxide, like a carbonyl group, activates an adjacent C-H leading to carbanion formation. However, the sulfur of an alkyl sulfide will also stabilize an adjacent carbanion. A sulfoxide behaves in many ways as $>S^+-O^-$. Sulfur compounds also undergo a number of rearrangements. This diversity of reaction pathways leads to a range of different structural types that may arise from quite simple substrates.

Much of the chemistry of the *Alliums* is concerned with the interaction between a sulfur, a sulfoxide or a sulfenic acid and an adjacent alkene. The alkene may be either in the 2-position allylic to the sulfur or in the 1-position as in a vinylsulfoxide. Clearly these have different reactive possibilities.

Garlic (*Allium sativum*) produces the amino acid alliin (prop-2-enyl-L-cysteine sulfoxide) **3.53** whilst onions (*A. cepa*) produce isoalliin which is the *E* (trans) geometric isomer of prop-1-enyl-L-cysteine sulfoxide **3.54**. The methyl, ethyl, propyl, butyl and butenyl-L-cysteine sulfoxides have also been identified amongst the flavouring constituents of various *Allium* species. Although both alliin and isoalliin have the same absolute stereochemistry for the sulfoxide, the operation of the Cahn–Ingold–Prelog sequence rules means that in alliin, the sulfoxide is designated as *S*, whilst in isoalliin, it is assigned the *R* configuration. The chemistry of the *Alliums* is dominated by the initial rapid cleavage of these alkylated cysteine sulfoxides to sulfenic acids (RSOH) mediated by the enzyme system, alliinase. The sulfenic acids then undergo a series of secondary reactions which depend upon their structure.

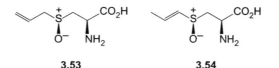

3.53   **3.54**

3.7 ONIONS

The chemistry of the onion, *Allium cepa*, has been studied for many years because of the organoleptic and beneficial properties of its constituents. Extraction and distillation of onion oils has given a range of sulfur compounds. These were first detected by Wertheim in 1844. Diallyldisulfide, propenylpropyldisulfide and dipropyldisulfide were reported by Semmler in 1892.

Onions contain the amino acids (*E*)-prop-1-enylcysteine sulfoxide (isoalliin) **3.54**, propyl-L-cysteinesulfoxide and *S*-methylcysteine sulfoxide. These amino acids, which can amount to as much as 1% of the dry weight of the onion, are stored as the γ-glutamyl derivatives. The enzyme system alliinase is responsible for the formation of the volatile metabolites from which, amongst other products, the onion lachrymatory factor **3.55** is generated by the action of a second enzyme. Whereas the amino acids are located in the storage mesophyll cell compartments of the onion, the enzyme system is held in the vascular bundle sheath cells. Only when the bulb is damaged and these separate compartments are ruptured, do the enzyme system and its substrate come into contact and fragmentation occurs to form prop-1-enylsulfenic acid **3.56**. Alliinase is a pyridoxal-dependent enzyme which depends for its activity on the formation of imines from the amino acid. This facilitates the fragmentation of the prop-1-enylcysteine sulfoxide to form the sulfenic acid. A number of these very reactive sulfenic acids, which have a half life of less than a second at room temperature, have been detected by the application of direct analysis in real time mass spectrometry (DART-MS) to *Allium* chemistry.

Prop-1-enylsulfenic acid **3.56** is the substrate for the onion lachrymatory factor synthase which catalyzes a rearrangement to form the powerful lachrymator *E*-thiopropanal *S*-oxide **3.55**. The *E* geometry of the thioaldehyde *S*-oxide was established through the formation of Diels–Alder adducts and is in accord with an electrocyclic mechanism for its biosynthesis. The genes which code for the formation of the lachrymatory factor synthase have been deleted to create a genetically modified 'tearless' onion.

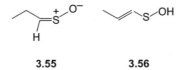

3.55   **3.56**

Prop-1-enylsulfenic acid **3.56** can undergo a series of dimerization and electrocyclic reactions to generate the *cis* **3.57** and *trans* **3.58** zwiebelanes (the *cis* and *trans* refer to the relative geometry of the methyl groups). The initial dimerization to form **3.59** involves a loss of water which is followed by an electrocyclic and rearrangement sequence. These compounds inhibit thromboxane biosynthesis in platelets. The thromboxanes are hormones which mediate vasoconstriction, platelet aggregation and clot formation. There is a formal similarity between the zwiebelanes and the oxygen bridged ring system of the thromboxanes. A related compound is onionin A which is the S-oxide of 3,4-dimethyl-5(*E*-prop-1-enyl)-tetrahydrothiophen-2-sulfoxide **3.60**. It has been shown to suppress tumour cell proliferation.

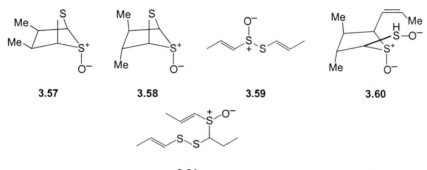

| 3.57 | 3.58 | 3.59 | 3.60 |

3.61

Another group of reactions leads to the cepaenes, *e.g.* **3.61**, which also have antithrombotic activity and are lipoxygenase inhibitors. The lipoxygenases are iron-containing enzymes which are involved in the oxygenation of unsaturated fatty acids to form hydroperoxides and their decomposition products. The formation of the cepaenes is based on the reaction of thiopropanal S-oxide with sulfenic acids to give an intermediate containing an ethyl group. A second reaction with a sulfenic acid then generates the cepaenes.

The available sulfur content of the soil, as sulfate and in the lower oxidation levels of sulfur, is an important determinant in the production of these compounds. There was a custom of putting soot from coal fires around onions. Trapped sulfur in the soot probably provided a useful supplement to natural sulfates, as well as having an antimicrobial activity.

The volatile sulfur metabolites, particularly dipropyldisulfide and propanethiol are oviposition stimulants for the onion fly, *Delia antigua*. The larvae of this pest cause considerable damage to the onion. The thiols are also released when the onion is damaged by soft rot bacteria. The fungus *Sclerotinia cepivorum* produces an onion white rot and is a serious soil-borne pathogen. A number of microbial biocontrol agents based on *Trichoderma* species and *Coniothyrium minitans* have been developed. 6-n-Pentylpyrone **3.62**, which is a metabolite of *Trichoderma harzianum*, inhibits the growth of *S. cepivorum*.

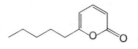

3.62

Onions are also an important dietary source of flavonoids. The major flavonol is quercetin **3.63** which occurs as the aglycone. Quercetin was obtained from onion skins by Perkin and Hummel in 1896. The outer skin of onion bulbs has been used for dyeing textiles and even for colouring egg shells at Easter. The amounts of the flavonols vary widely with the cultivar and with the conditions of growth. Yellow onions contain 270–1190 mg of flavonol per kg (fresh weight), whilst red onions contain 417–1917 mg of flavonol per kg. Quercetin 4′-glucoside and 3,4′-diglucoside are the major components, whilst other glucosides including those of kaempferol **3.64** and isorhamnetin **3.65** have been reported. Onions are dried for a few weeks before being stored to prevent post-harvest attack by the fungus *Botrytis allii*. Changes take place during that time and a yellow xanthylium pigment is formed. This pigment is known as cepaic acid **3.66**. It is formed from the reaction of glyoxylic acid and two molecules of phloroglucinol, which are derived from ring A of quercetin.

The anthocyanins of red onions are predominantly cyanidin **3.67** glucosides, a number of which are also acylated with malonic acid. The amounts which are present vary between 39–240 mg per kg (fresh weight). Smaller amounts of dihydroflavonols including taxifolin **3.68** have also been isolated from various cultivars.

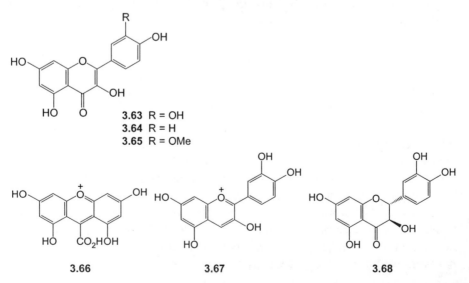

3.63 R = OH
3.64 R = H
3.65 R = OMe

3.66   **3.67**   **3.68**

The presence of flavonoids in onions has been associated with some of their beneficial effects on age-related diseases. They are powerful antioxidants but their medical effects may also arise from other properties such as their anti-inflammatory activity. There are some indications that these compounds may have an anti-proliferative effect on colon cancer cells.

Analysis of red onions has shown that apart from flavonoids, they also contain some steroidal furostanol saponins, known as the tropeosides **3.69** and ascalonicosides. The latter were also obtained from shallots (*A. ascalonicum*) which are closely related to onions. These saponins show antispasmodic activity which may explain the traditional use of onions in the treatment of gastrointestinal problems. A similar group of steroidal furostanol saponins, known as the fistulosaponins A–F, have been obtained from *Allium fistulosum* (Welsh onions or Spring onions).

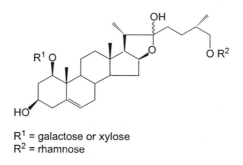

R^1 = galactose or xylose
R^2 = rhamnose

3.69

3.8 GARLIC

Many of the health benefits of garlic (*Allium sativum*) can be attributed to its phytochemistry and in particular to the compounds containing sulfur which are present. The amino acids based on cysteine are very similar to those found in onions. However, there is a subtle difference with considerable chemical consequences. Garlic contains prop-2-enyl-L-cysteine sulfoxide (alliin) **3.53** rather than the prop-1-enyl isomer. The enzyme alliinase converts this to an allylsulfenic acid which undergoes a rapid dimerization to the thiosulfinate, allicin **3.70**. This compound has antibacterial properties and is reduced to diallyldisulfide. A number of these compounds inhibit the growth of *Helicobacter pylori* which is the causative organism for stomach ulcers.

Allicin undergoes some interesting chemistry. The oxygen of the sulfoxide can act as an internal base allowing allicin to fragment to thioacrolein **3.71** and a propenyl sulfenic acid **3.72**. Thioalkylation of a second molecule of allicin **3.70** by the sulfenic acid **3.72** generates a sulfonium ion **3.73**, which undergoes a second fragmentation to generate an unsaturated sulfonium compound **3.74**. This in turn can add propenyl sulfenic acid **3.72** to give ajoene **3.75**. This product, which exists in *E* and *Z* geometrical isomers, occurs in garlic and possesses powerful antithrombotic activity and inhibits platelet aggregation. The thioacrolein **3.71** which is formed in this sequence undergoes various dimerization reactions to give cyclic disulfides. Thiophene derivatives have also been found.

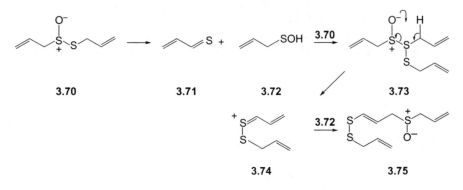

The sulfur content of 'garlic breath' includes allylmethylsulfide, allylmethyldisulfide, diallylsulfide and disulfide, prop-2-enethiol and hydrogen sulfide. Some selenium analogues have also been detected and are derived from selenocysteine.

3.9 LEEKS

The phytochemistry of other edible *Allium* species such as the leek, *Allium porrum* is also dominated by sulfur. In contrast to onions and garlic where there is one major cysteine derivative, either isoalliin or alliin present, leeks contain three compounds: methiin, propiin and isoalliin. The enzyme systems alliinase and the lachrymatory factor synthase are also present. Consequently, a complex mixture of sulfur-containing volatiles is produced on crushing leeks. The edible portion of the leek, once it is cooked, has a relatively mild taste compared with onions. Leeks also contain a group of spirostanol sapogenins, known as the porrigenins, *e.g.* **3.76**. These show antiproliferative activity against various cancer cell lines. In addition, leeks produce a number of dibenzofurans with antifungal activity against *Fusarium* species.

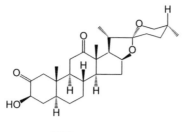

3.76

3.10 ASPARAGUS

Asparagus (*Asparagus officinalis*) is a popular seasonal delicacy which has been cultivated since Roman times. This vegetable was classified as a member of the Liliaceae family which included the Alliums but it has since been moved to the Asparagaceae. The plant is a perennial and it is the young shoots that are harvested in the spring. The amino acid, asparagine **3.77** which is the amide of aspartic acid, was first isolated in 1806 from asparagus juice. Like glutamine, it plays an important role in nitrogen metabolism in plants. Pharmacological studies have demonstrated that asparagus has anti-inflammatory, cytotoxic and diuretic activity, whilst nutritional studies indicate that it is a good source of folic acid, vitamin C and potassium ions.

A relationship to the Alliums is indicated by the presence of a number of sulfur-containing compounds derived from cysteine. These include *S*-(2-carboxy-n-propyl)- L-cysteine **3.78** and some disulfides which contribute to the flavour. The unusual disulfide, asparagusic acid **3.79**, is found in both the roots and edible portions of asparagus. It is biosynthesized from isobutyric acid *via* methacrylic acid, 2-methyl-3-mercaptopropionic acid and S-(2-carboxy-n-propyl)cysteine. It is a plant growth inhibitor and it has a powerful nematocidal activity, accounting for the recommendation to grow asparagus as a companion plant to other crops that suffer attack from nematodes. The methyl esters of the epimeric sulfoxides of asparagusic acid, in which the S-oxide takes up either a *syn* or *anti* configuration relative to the methyl ester, have been found in asparagus. The unpleasant smell of urine after eating asparagus probably arises from methanethiol, dimethyldisulfide and dimethylsulfoxide, and other metabolites of asparagusic acid.

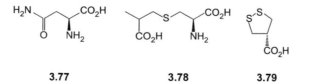

3.77	**3.78**	**3.79**

The aerial parts of asparagus also contain some acetylenes, such as asparenyne **3.80** and its 2-hydroxy derivative, together with some glycerol esters such as 1,3-O-di-*p*-coumaroylglycerol. The carotenoid degradation product, blumenol C **3.81**, has also been isolated. The roots of the asparagus plant have been shown to contain steroidal saponins, in which sarsasapogenin **3.82** and its 17α-hydroxy derivatives are the aglycones. The roots also contain allelopathic phytotoxic constituents, such as ferulic acid and methylenedioxycinnamic acid.

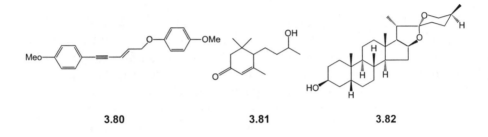

| 3.80 | 3.81 | 3.82 |

3.11   CELERY

Celery (*Apium graveolens*) belongs to the Apiaceae (Umbelliferae) which is the same family of plants as the carrot, parsnip, parsley and even the giant hogweed. Not unexpectedly, there are some similarities between their chemistry. Wild celery can be found growing in damp localities in Britain and it is quite strong smelling.

The edible stalks of celery are a structural part of the plant involved in supporting the leaves and in transporting water and minerals to them, and in returning the products of photosynthesis back to the roots. Celeriac is the fleshy bulbous root of a variant (*A graveolens* var. *rapaceum*). An oil (celery seed oil), which is used principally as a flavouring, is obtained from the fruit (seeds) of celery. The celery that is sold is 'blanched' or 'trenched' to protect the stems from sunlight, so that less chlorophyll and some other compounds typical of this family are produced.

A number of uses of celery are reported in folk medicine. The juice is alleged to alleviate some rheumatoid conditions and to be mildly diuretic with a beneficial effect on the kidneys. This applies particularly to the root extract. There is at least one report of it being an aphrodisiac, and other reports of one of the constituents leading to an abortive action. The relatively high sodium and potassium ion content, along with the presence of the flavone glycoside, apiin **3.83**, may be associated with the diuretic action.

The flavouring constituents which are found more especially in celery seed oil include the lactones, 3-n-butylphthalide **3.84**, sedanenolide **3.85** and derivatives of the phenylpropanoid, apiole **3.86**, as well as mono-terpenes related to limonene and sesquiterpenes (celerosides) related to β-selinene **3.87**. These relatively unusual constituents confer an attractive taste to celery. The lactones are reported to have some insecticidal properties including activity against mosquito larvae, together with some antifungal activity. 3-n-Butylphthalide is reported to help blood vessels dilate and so reduce blood pressure. The seed oil, like that of other members of the Apiaceae contains an unusual unsaturated fatty acid, petroselinic acid (*Z-6*-octadecenoic acid). The polyacetylenes, falcarinol **3.37** and falcarindiol **3.38**, which are also found in parsnips and carrots, have been detected in celery bulbs. These compounds have antifungal and antibacterial, as well as cytotoxic activity. The penta-cyclic triterpene, 11,21-dioxo-2β,3β,15α-trihydroxyurs-12-ene-2-O-β-D-glucopyranoside **3.88**, and some relatives have been isolated from celery. Triterpenes of this type have anti-inflammatory activity.

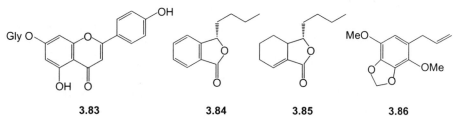

3.83   **3.84**   **3.85**   **3.86**

3.87

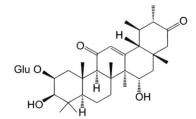

3.88

There is a negative side to the constituents of celery. Like other members of the Apiaceae such as the parsnip, celery produces furanocoumarins such as psoralen **3.39**, its 5-methoxy derivative bergapten and xanthotoxin **3.40**. These compounds are phytoalexins and their production is stimulated as a consequence of fungal attack. When celery is infected by, for example, *Fusarium oxysporum* var. *apii*, these compounds tend to spread throughout the stem. However, these furanocoumarins are phototoxic to man and can produce a reddening of the skin when exposed to strong sunlight. There is a report of a person who had consumed a large quantity of celery and then spent half an hour on a sunbed, ending up in hospital with severe erythema. The concentrations of these furanocoumarins is highest in the older outer leaves (*ca.* 45 μg per g fresh weight) and least in the immature inner heart and petioles (1–3 μg per g). Care needs to be taken when trimming off the outer coarse leaves of celery to avoid excessive contact with the plant and the consequent photodermatitis.

3.12 RHUBARB

The species of rhubarb (*Rheum rhabarbarum*, *R. rhaponticum* and *R. x hybridum*) are strictly vegetables, although the plants are grown for their leaf stalks (petioles) which are used as substitutes for fruit, particularly in the spring. Rhubarb is a perennial plant which, in temperate climates, dies back in the autumn and grows vigorously in the spring, often under sheltered forcing conditions, to provide an early crop. The medicinal properties of the roots and rhizomes were highly valued in the 18th and 19th century when much of the plant was imported from China. Today it is widely grown in the UK and it is the stalk which is used as a foodstuff.

Rhubarb is quite acidic containing malic, citric and oxalic acids. The latter is quite poisonous (LD_{50} in dogs $= 1.0$ g per kg). However, the highest concentration of oxalic acid, which is in the leaves, is 0.6% and hence rather a large quantity would need to be eaten. Nevertheless, lower amounts can cause gastric disturbance and renal damage. There are reports that in the past, rhubarb leaves were cooked as a substitute for a green vegetable. Presumably the oxalic acid was eluted on cooking.

The stalk contains 10–20 mg per 100 g of vitamin C. The red colour of the stalk arises from a mixture of the anthocyanin pigments, cyanidin **3.68** 3-glucoside and 3-rutinoside. The polyhydroxyanthraquinones, chrysophanic acid **3.89**, rhein **3.90**, emodin **3.91** and aloe-emodin **3.92**, together with their monomethyl ethers are responsible for the astringent taste of rhubarb. This, together with the tartness arising from its acidity,

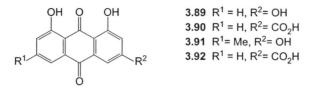

3.89 R^1 = H, R^2= OH
3.90 R^1 = H, R^2= CO_2H
3.91 R^1= Me, R^2= OH
3.92 R^1 = H, R^2= CO_2H

means that rhubarb requires sweetening with sugar and hence it did not become popular as a food until sugar was readily available. It is often used with other fruit in the spring and early summer as they become available. The purgative effect of rhubarb is associated with the sennosides **3.93** and with the glycosides of chrysophanic acid and emodin. Sennosides are metabolized by the intestinal bacteria to rhein anthrone.

The constituents of the roots and rhizomes of the Korean rhubarb, *R. undulatum*, which is used in Japanese, Chinese and Korean traditional medicine, have been thoroughly examined. The principal biologically active constituents have been shown to be stilbenes, such as rhaponti-genin **3.94**, resveratrol **3.95** and piceatannol **3.96**. They inhibit the production of nitric oxide from arginine and may have beneficial effects against various types of inflammation.

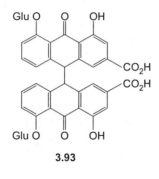

3.93

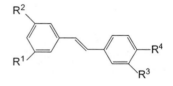

3.94 R^1= R^2= R^3= OH, R^4= OMe
3.95 R^1= R^2= R^4= OH, R^3= H
3.96 R^1= R^2= R^3= R^4= OH

The description of the constituents of root, bulb and stem vegetables in this chapter has been restricted to those that are commonly grown in the UK. Nevertheless, the fact that they belong to different families of plants is reflected in the structural diversity of their constituents and consequently the range of their biological activity and potential benefit to man.

CHAPTER 4

Green Leaf Vegetables

4.1 INTRODUCTION

The aerial parts of green vegetables have provided staple foodstuffs for many millenia. They contain a range of interesting and useful natural products which contribute to the beneficial and organoleptic properties of these plants. In this chapter, we will consider the cruciferous vegetables that belong to the Brassicaceae such as the cabbage, as well as greenstuffs such as spinach and lettuce which belong to other families. In some of these plants, the edible part is not the leaf but the flower or seed. Other members of these families have already been encountered in the previous chapter. Thus the turnip is the swollen root of *Brassica rapa* and the swede is that of *B. napus*.

Many of the Brassicas are closely related. Indeed the cabbage, broccoli, kale and brussel sprouts are all cultivars of *Brassica oleracea*, whilst there are close genetic relationships between *B. oleracea* and *B. juncea* (Oriental mustards), *B. rapa* (turnips) and *B. napus* (swede), as well as with the hybrid oil-seed rape and canola. Black mustard seeds, which are the source of the condiment, are usually obtained from *B. nigra*, whilst the mustard of 'mustard and cress' is *B. alba* (*Sinapis alba*). Not surprisingly, there is a considerable overlap between the chemistry of these species. As suggested by the botanical name *B. oleracea* and the application of oil-seed rape, parts of these plants have a high oil content. The oil from oil-seed rape contains the glycerides of oleic, linoleic, linolenic and palmitic acids. Elaidic acid is the *trans* isomer of oleic acid and because it is harmful, its formation has been bred out of modern

Chemistry in the Kitchen Garden
By James R. Hanson
© James R. Hanson 2011
Published by the Royal Society of Chemistry, www.rsc.org

cultivars. Canola (Canadian oil, low acid) is a cultivar of rape seed which was developed in Canada in the 1970s to produce an edible seed oil which was low in Elaidic acid. It is now grown quite widely in North America.

4.2 CHLOROPHYLL

The chloroplasts in the green leaves of plants contain the chlorophyll and carotenoid pigments which are involved in photosynthesis. Indeed spinach leaves are often used as a source of chlorophyll for student chromatography experiments. Chlorophyll 'a' **4.1** absorbs light at wavelengths between 400 and 450 nm and also between 650–700 nm, whilst the aldehyde, chlorophyll 'b' **4.2**, has absorption between 450–500 nm and also between 600–650 nm. The carotenoids and xanthophylls which are present absorb between 400 and 530 nm. However, none of these have significant absorption in the green region of the spectrum, hence the colour of the leaves.

Photosynthesis involves trapping the energy of sunlight by the reduction of atmospheric carbon dioxide to sugars using the hydrogen of water and releasing oxygen. Without the energy held by the sugars and then released during their catabolism, cellular metabolism and plant development would cease. Photosynthesis involves a series of light-dependent reactions to convert the light energy into ATP and NADPH. The second set of light-independent reactions utilize the ATP and NADPH to convert the carbon dioxide into sugars.

There are broad similarities between the biosynthesis of the tetra-pyrrole ring system of chlorophyll and that of the heme pigments. The tetrapyrrole porphyrin ring of uro'gen-III **4.3**, which is common to both pathways, is biosynthesized from four molecules of a pyrrole, porpho-bilinogen **4.4**. This is in turn obtained by the condensation of two molecules of δ-aminolevulinic acid ($H_2NCH_2.CO.CH_2CH_2CO_2H$). However, in the biosynthesis of the heme pigments in animals and bacteria, this is formed from glycine and succinyl co-enzyme A, whilst in plants the C_5 skeleton of δ-aminolevulinic acid is obtained from the intact skeleton of glutamic acid. A linear 1-hydroxymethylbilane arises from four molecules of porphobilinogen. The formation of the cyclic uro'gen III involves a rearrangement of ring D, which has the effect of interchanging a C_2 and a C_3 side chain. The extended conjugation and the magnesium are introduced at this protoporphyrin stage.

The last stages of the biosynthesis of chlorophyll involve the light-dependent reduction of ring D of protochlorophyllide to chlorophyll by the enzyme, protochlorophyllide reductase. In the absence of light, this step does not occur and the plant does not turn green. Thus the leaves in the interior of the cabbage and those that are used in the preparation of coleslaw are white. Spectroscopic studies have shown that the chlorophyll levels gradually decrease on going from the outer leaves to the inner leaves of cabbage where protochlorophyllide and other protochlorophylls dominate. However, these leaves may contain the carotenoid xanthophylls such as lutein **4.5** and lutein epoxide whose biosynthesis is not light dependent. The blanching of celery by trenching or surrounding the stems of the plant with a tube, also involves the exclusion of light to inhibit this last step.

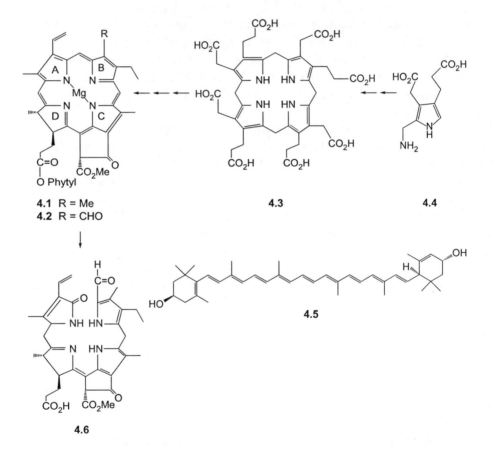

4.1 R = Me
4.2 R = CHO

4.3

4.4

4.6

Chlorophyll is degraded quite rapidly. There is a continuous process of degradation and resynthesis in the leaf. When a plant is under water stress and there is insufficient water for the biosynthesis and transport of sugars, chlorophyll synthesis ceases and the leaves turn yellow. The central magnesium is also quite labile. Its removal gives the pheophytins a and b which possess a brownish colour. When green vegetables are cooked in boiling water, the magnesium may be removed from the chlorophyll, particularly if the water is slightly acidic. Cooked cabbage can loose its colour becoming quite pale or even slightly brown in colour. The addition of small amounts of sodium bicarbonate to neutralize the plant acids can reduce colour loss. This loss of colour is a problem in canning peas. The green colour has sometimes been supplemented by the addition of sodium copper chlorophyllin in which the copper complex is more stable.

The catabolism of chlorophyll involves the loss of the magnesium and scission of the chlorin ring system. Whilst some of the early catabolites are fluorescent, the majority are non-fluorescent. These metabolites are exemplified by **4.6** which has a rust colour and contributes, along with the carotenoids, to the brown colour of senescent leaves.

4.3 GLUCOSINOLATES

The glucosinolates with the general formula **4.7** are characteristic constituents of the Brassicaceae. Much of the biological activity of the glucosinolates can be attributed to their enzymatic hydrolysis and rearrangement products which lead to the formation of isothiocyanates **4.8**. The hydrolysis and rearrangement, a biological equivalent of a Beckmann rearrangement, is catalyzed by the enzyme myrosinase. This enzyme system is present in separate compartments in the plants to the glucosinolates, and only when the tissues are disrupted can the enzyme and substrate come into contact to afford the isothiocyanate. This has implications for the interaction between the plants and herbivorous insects and fungal pathogens. The isothiocyanates, which are sometimes known as mustard oils, are more volatile than the parent glucosinolate and may behave as deterrents for some species and attractants for others. These isothiocyanates react readily with nucleophiles such as the sulfur of cysteine to give cysteine, glutathione or protein conjugates and with water to give thiocarbamates. Studies on blood plasma indicate that the majority of dietary glucosinolates can be converted to isothiocyanates by bacteria in the gastrointestinal tract. Glucosinolate hydrolysis products from the decay of Brassica plant tissues can inhibit the growth of nitrification bacteria in the soil and so diminish the loss of nitrogen from the soil by this part of the nitrogen cycle. They have also been shown to act as allelochemicals by inhibiting the germination of weed seeds and the development of soil-borne pests.

There is also an important implication in the preparation of these vegetables for the table. Like many enzyme systems, myrosinase is thermally labile and if it is destroyed before the glucosinolates undergo enzymatic hydrolysis, a different chemical degradation can occur leading to the unpleasant odours associated with cooking 'greens'. The varying extents of enzymatic and chemical degradation of glucosinolates may account in part for differences between chopped cabbage products, such as coleslaw and cooked cabbage, and even between frozen, blanched and fresh brussel sprouts. There is an interesting correlation which may be genetic, between people who can detect phenyl thiocarbamide and find it bitter and unpleasant and those who dislike cruciferous vegetables. This dislike may be related to the decomposition products of isothiocyanates from these vegetables. Brussel sprouts have a certain notoriety in this connection.

There can be quite wide variations in the structure of the individual glucosinolates (**4.9–4.16**) and the amounts that are present in different species. There are even variations between cultivars. The seeds of Brassicas contain glucosinolates, such as sinigrin **4.9** and glucosinalbin **4.12**, as protective agents. The oil which is used in the preparation of mustard is obtained by the steam distillation of the seeds of *B. nigra* or *B. juncea*. Hence the alternative mustard oil name for the glucosinolates. Sinapic acid (3,5-dimethoxy-4-hydroxycinnamic acid) and its choline ester, sinapine, are also found in black mustard seed. Radishes (*Raphanus sativus*), described in the previous chapter, are also members of this family and owe their pungent taste to the presence of glucosinolates. Rocket (*Eruca vesicaria*) is also a member of the Brassicaceae family. Its leaves may be added to a salad to give a peppery taste which has been attributed to the presence of glucosinolates. The production of glucosinolates by the Brassicas requires the presence of sulfur in the soil and this may need to be supplemented by the addition of a sulfate such as magnesium or ammonium sulfate.

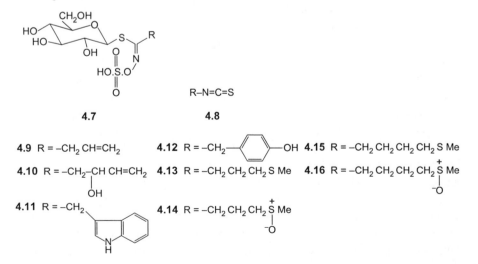

4.7   **4.8**

4.9 R = –CH$_2$CH=CH$_2$ **4.12** R = –CH$_2$—⟨benzene⟩—OH **4.15** R = –CH$_2$CH$_2$CH$_2$CH$_2$S Me

4.10 R = –CH$_2$–CH CH=CH$_2$ **4.13** R = –CH$_2$CH$_2$CH$_2$S Me   **4.16** R = –CH$_2$CH$_2$CH$_2$CH$_2$ $\overset{+}{S}$ Me
            |
            OH $^-$O

4.11 R = –CH$_2$⟨indole⟩   **4.14** R = –CH$_2$CH$_2$CH$_2$ $\overset{+}{S}$ Me
                                           $^-$O

4.4 PHYTOALEXINS OF THE BRASSICAS

The phytoalexins of the Brassicas have been the subject of a number of investigations. The initiation of the biosynthesis of the phytoalexins, as opposed to the phytoanticipins, requires an elicitor such as a microbial or chemical stress. In the case of the Brassicaceae family most of the compounds are indoles derived from L-tryptophan many of which, such as brassinin **4.17** from the cabbage (*Brassica oleraceae* varn. *capitata*), also contain sulfur. Brussel sprouts (*B. oleraceae* varn. *gemmifera*) produce a thiolcarbamate, brussalexin A **4.18**. Structural analogies suggest that the biosynthesis of these compounds may involve an indolylglucosinolate pathway, although the stage at which divergence occurs is not yet clear. The sulfur comes from the amino acid cysteine. The susceptibility of these plants to fungal diseases may be related to the ability of virulent phytopathogens such as *Rhizoctonia solani, Sclerotinia sclerotiorum* and *Alternaria brassicicola* to metabolize the phytoalexins. Thus brassinin is hydrolysed to an amine and then converted to indole-3-carboxaldehyde, whilst the nitrogen of the indole is glucosylated.

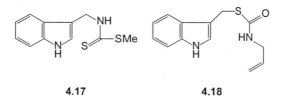

4.17   **4.18**

Leafy vegetables are subject to potential photo-oxidative stress and consequently they contain protective antioxidants, such as the carotenoids, the tocopherols and vitamin C. Some cultivars, such as red cabbage, also contain anthocyanidins.

4.5 CABBAGE

Cabbage (*Brassica oleracea* varn. *capitita*) has been grown as a staple crop for many years. It contains many of the types of compound that are characteristic of the Brassicaceae. One of the glucosinolates which is found in the cabbage is sinigrin **4.9** and this gives rise to allylisothiocyanate **4.19** which contributes to the smell of cabbage. This isothiocyanate is a powerful antifungal agent which acts against the fungus *Peronospora parasitica* that forms a powdery mildew on the cabbage.

$$\text{N}=\text{C}=\text{S}$$

4.19

Sinigrin **4.9** is toxic to insects and acts as a feeding deterrent to many butterflies. However, the cabbage white butterfly, *Pieris brassicae,* can tolerate sinigrin and uses it as an oviposition stimulant. The absence of competition from other species provides the cabbage white caterpillar with an undisputed source of food. The parent butterfly has developed a further strategy to ensure a source of food for its progeny. It deposits oviposition deterrents on the eggs to deter another female from laying eggs in the vicinity. These compounds, 5-deoxymiriamide **4.20**, miriamide **4.21** and its 5-glycoside are caffeic acid amides of a hydroxlated anthranilic acid. When the caterpillars of the butterfly damage the cabbage leaf, volatile semiochemicals are released. These herbivore-induced plant volatiles include some terpenes, such as linalool and farnesene, which attract a parasitic wasp, *Cotesia rubecula.* The female of the wasp lays her eggs within the caterpillar. These eventually develop into the larvae of the wasp and kill the caterpillar. This parasitoid and a relative, *C. glomerata,* have been proposed as biocontrol agents. The white pigment of the wings of the cabbage white butterfly is a pterin, leucopterin **4.22**, whilst xanthopterin **4.23** is the pigment of the yellow brimstone butterfly.

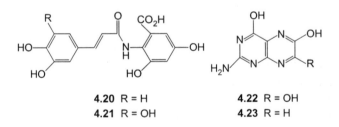

4.20 R = H   **4.22** R = OH
4.21 R = OH   **4.23** R = H

The cabbage aphid, *Brevicoryne brassicae,* stores sinigrin and myrosinase and releases allylisothiocyanate when it is consumed by a ladybird, thus deterring further attack. The glucosinolates are also oviposition stimulants for the cabbage root fly, the larvae of which tunnel into the roots of many of the Brassicas. The indoleglucosinolates and the triazafluorene **4.24**, derived from tryptamine and cysteine, stimulate oviposition by the female, *Delia radicum* in the soil beneath the plant. This reveals a reason why placing a piece of cardboard on the soil around the stem of a young cabbage provides protection against this pest, since it prevents the eggs from developing and the larvae from reaching the roots.

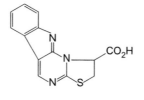

4.24

A glucosinolate which can be the cause of a thyroid disease is progoitrin **4.10**. This occurs in a number of Brassicas including cabbage. When this is cleaved by myrosinase, the isothiocyanate which is released cyclizes to form the toxic oxazolidinethione, goitrin **4.25**. Some white cabbage cultivars produce as much as 20 mg per kg in the autumn. Goitrin inhibits the production of the hormones, triiodothyronine and thyroxine, by the thyroid gland. The thyroid gland may enlarge to compensate for this leading to 'cabbage goitre'.

Cabbages produce several indoles. Glucobrassicin **4.11** is a glucosinolate which contains a 3-indolylmethyl residue. Glucobrassicin and its 4-hydroxy and 4-methoxy derivatives are converted to phytoalexins, such as brassinin **4.17**, cyclobrassinin **4.26** and brassicanal C **4.27**, in response to microbial attack. The unusual dithiomethyl ethers, methoxybrassenins A **4.28** and B **4.29**, have also been isolated as stress metabolites from cabbage inoculated with *Pseudomonas cichorii*. Brassicanal C **4.27** possesses a chiral sulfinate and its absolute configuration has been shown to be *S*. These indoles, along with the plant hormone, indolylacetic acid, can be the source of indole-3-aldehyde and 3-carbinol which have been detected in amounts of about 0.02% in the leaves. A valuable compound which is derived from these is 3,3'-diindolylmethane **4.30**. This compound possesses tumour inhibitory activity. It has been shown to inhibit the proliferation of hormone-dependent breast and prostate cancer cells. In the case of human prostate cancer cells, it is an androgen receptor antagonist. The indolylmethylisothiocyanates are unstable and release the thiocyanate ion and a reactive carbonium ion in a fragmentation reaction which is facilitated by the indolic nitrogen. The products of reaction of this carbonium ion depend on the nucleophiles that are available. Apart from the formation of the carbinol, it will also react with vitamin C to form an ascorbigen. Reaction with allyl isothiocyanate, produced from sinigrin, could generate an unusual thiocarbamate, the phytoalexin brussalexin A **4.18**, which has been isolated from brussel sprouts (*B. oleracea* varn. *gummifera*).

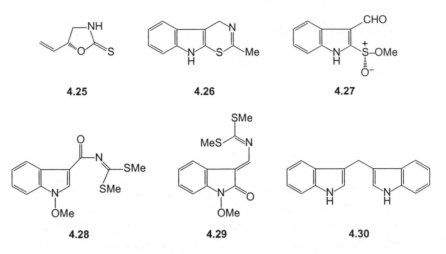

4.25   **4.26**   **4.27**

4.28   **4.29**   **4.30**

Various other groups of compound have been found in cabbages including vitamin C, β-carotene and the antioxidant, α-tocopherol. The leaf waxes include long-chain hydrocarbons and the ketones, 14- and 15-nonacosanone. The pigments of red cabbage are cyanidin **4.31** 3,5-diglucoside and the cyanidin 3-sophoroside 5-glucosides. These are also acylated with 1–2 moles of sinapic acid. This red pigment is pH sensitive, turning blue under alkaline conditions. Hence red cabbage should be cooked with a little vinegar to retain its red colour.

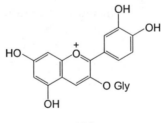

4.31

4.6 BROCCOLI

Two types of broccoli are commonly encountered. The calabrese variety which produces large green heads and is typically grown commercially, and the purple sprouting variety. The florets of broccoli (*B. oleracea* varn. *botrytis caput*) when compared to cabbage contain relatively high levels of β-carotene (0.89 mg per 100 g fresh weight), α-tocopherol (1.62 mg per 100 g) and vitamin C (74.7 mg per 100 g). As harvested broccoli ages, the chlorophyll decomposes first and the yellow to orange carotenoids remain. One of the major glucosinolates that is present in broccoli is glucoraphanin (4-methylsulfinylbutylglucosinolate) **4.16**.

When this is cleaved by myrosinase it gives rise to sulforaphane (4-methylsulfinylbutylisothiocyanate) **4.32**. This compound has several beneficial properties. It inhibits the growth of the bacterium *Helicobacter pylori,* which is associated with the formation of ulcers in the stomach. It also induces the phase two enzymes in the liver which detoxify metabolites produced by the oxidative phase one metabolism by linking them with glutathione. There is also an indication that this compound may function to protect the cardiovascular system. Together with its reduction product, erucin, it can enhance the biological activity of the transforming growth factor-β which plays a role in restraining excessive mammalian cell growth. The regular consumption of broccoli has been linked to reduced low density lipoprotein and total cholesterol levels. Sulforaphane **4.32** and the related compounds 5-methylsulfinylpentylnitrile and 4-methylsulfinylbutylnitrile have also been isolated from rapidly growing broccoli seed sprouts.

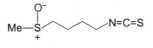

4.32

4.7 CAULIFLOWER

Cauliflower, which is closely related to broccoli, is a variant of the *B. oleraceae, Botrytis group*. The commonly used parts of the plant are the florets. In common with other Brassicas, it contains glucosinolates such as glucoiberverin **4.13** and its sulfoxide, glucoiberin **4.14**, as well as the homologues, glucoerucin and glucoraphanin. These compounds by affording, for example, sulforaphane, confer beneficial properties to cauliflower. Indole-3-carbinol has also been reported. The characteristic flavour and odour of cooked cauliflower are associated with the decomposition products of these glucosinolates. These include alkyl cyanides such as 4-(methylthio)butyl cyanide and 4-(methylthio)butylisothiocyanate from glucoerucin **4.15** and 3-(methylthio)propylcyanide and isothiocyanate from glucoibeverin **4.13**. Various dimethylpolysulfides and compounds such as hex-3-en-l-ol, which arise from degradation of fatty acids, have also been detected. Although the common form of cauliflower has a white head, pigmented forms are sometimes grown. These include an orange, a purple and a green form. The more spiky head forms of the latter are known as Romanesco varieties. The orange pigmentation arises from β-carotene **4.33** and is a good source of vitamin A, whilst the purple form owes its pigmentation to the presence of antioxidant anthocyanins, particularly *p*-coumaryl and feruloyl esters of cyanidin **4.31** 3-sophoroside-5-glucoside. Not suprisingly, bearing in mind their water solubility, cooking leads to a decrease in the anthocyanin content.

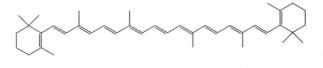

4.33

4.8   SPINACH

Spinach (*Spinacia oleracea*) belongs to the family Amaranthaceae, which was formerly known as the Chenopodiaceae. The leaves of this plant have a useful nutritional value. Spinach is not only a rich source of chlorophyll but also of carotenoids. It is a good source of vitamin C and a number of metal ions including calcium and iron. However, the old story concerning its very high iron content on which the Popeye advertisments were based is in error. It has even been suggested that a decimal point was in the wrong place in the original analysis. Later determinations suggest that the available iron is in the region of 3–4 mg per 100 g fresh weight. Spinach produces quite high levels of oxalic acid both by degradation of ascorbic acid and from glyoxalic acid. This can affect the absorption of iron and calcium from the plant. Much of the oxalic acid, which can be as much as 740 mg per 100 g fresh weight in a summer cultivar, is removed in cooking. The precipitation of calcium oxalate can give rise to kidney stones. Spinach, like a number of leafy vegetables, can also absorb quite high levels of nitrate from the soil. This can be reduced to the harmful nitrite ion by the intestinal bacteria.

The total carotenoid content, mainly β-carotene **4.33**, lutein **4.5**, the 5,6;5′6′-diepoxide violaxanthin and the allene 9′-(*Z*)-neoxanthin, can be in the range 170–220 mg per kg. β-Carotene which is a significant component, is a precursor of vitamin A. Spinach produces an unusual series of flavonoid glycosides possessing an extra oxygen function at C-6. These glycosides are derived from the aglycones patuletin **4.34** (quercetagetin 6-methyl ether) and spinacetin (quercetagetin 6,3′-dimethyl ether). Some of these occur as their glucuronides rather than as their glucosides. Interestingly, several of these compounds have exhibited antimutagenic activity.

Spinach has proved to be a useful plant species in which to study a number of problems of plant biochemistry. The gibberellin plant hormones that are formed under different growing conditions have been examined by gas chromatography-mass spectrometry. Due to their low natural abundance, identification entailed the synthesis of the putative structures from more readily available fungal gibberellins such as gibberellic acid. The presence was established of sequences of C_{20} and C_{19} gibberellins from gibberellin A_{12} in which hydroxyl groups were introduced at C-13 and then at C-2β (*e.g.* GA_{97} 2β,13-dihydroxyGA_{12}) **4.35**. Spinach has also been one of the sources of a peptide, rubiscolin 6 (TyrProLeuAspLeuPhe), which targets the δ-opioid receptor and has an analgesic and anxiolytic effect.

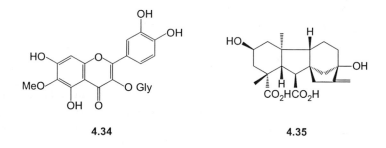

4.34 **4.35**

4.9 LETTUCE

Lettuce (*Lactuca sativa*) is widely grown as a salad vegetable. There are many varieties which range from the crisp 'Iceberg' lettuce to the 'Cos' lettuce, together with loose leaf varieties which are used on a 'cut and come again' basis. A number of related species, such as *L. virosa* and *L. tatarica*, have also been used as salad vegetables.

Typical of a member of the Asteraceae (Compositae) family, lettuce produces sesquiterpene lactones. In this case they are guaianolides including lactucin **4.36**, dihydrolactucin and the *p*-hydroxyphenylacetic acid ester of lactucin which is known as lactupicrin. Wounding the leaves and stems of *Lactuca* species releases a milky latex, hence the name of these plants. This contains the 15-oxalate and 8-sulfate esters of lactucin and other lactones. The sequiterpene lactones are quite bitter and consequently lettuce cultivars have been bred in which their concentration has been reduced. If a lettuce is allowed to bolt, the latex is often formed in the stem. However, these lactones play an important role as insect antifeedants. The related sesquiterpenoid lactone, lettucenin A **4.37**, is a phytoalexin whose formation is induced by lettuce downy mildew, *Bremia lactucae*.

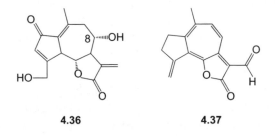

4.36   **4.37**

  Lactucarium (lettuce opium), which has mild sedative and euphoric properties, is the dried latex of the wild acrid lettuce, *L. virosa*, and the prickly lettuce, *L. serriola* (*L. scariola*). It contains lactucin and lactupicrin as the bioactive agents. In the past, these plants were grown as medicinal herbs. Cases of poisoning have been reported from overdoses.

  Some triterpenes including lupeol and β-amyrin and sterols such as β-sitosterol have been reported. Like many members of the Asteraceae, lettuce also produces some polyacetylenes including the C_{17} highly unsaturated hydrocarbon, *trans,trans*-heptadeca-8,10,16-triene-2,4,6,-triyne **4.38** which is probably a decarboxylation product of a C_{18} acid. The colouring matter of the red cultivars of lettuce is cyanidin **4.31** 3-(6″-malonylglucoside).

4.10 CHICORY

Chicory (*Chicorium intybus*) and endive (*C. endivia*) are related species which also belong to the Asteraceae. The leaves of chicory are edible whilst the deep tap-root when roasted and ground, is sometimes blended with coffee. It contains the bitter sesquiterpenoid lactones, lactucin **4.36** and lactucopicrin, as well as 8-deoxylactucin. They are reported to have a bitterness which is of the same order of magnitude as quinine and to be detectable at concentrations of 2 ppm. They protect the plant against attack by herbivores. The pigments of these plants include cyanidin **4.31** 3-malonylglucoside, luteolin **4.39** 7-glucuronide, together with quercetin **4.40** and kaempferol **4.41** 3-glucuronides. Some coumarins such as 6,7-dihydroxycoumarin (cichoriin) have also been isolated. The cichoric acids are the caffeic acid esters of tartaric acid and act as insect antifeedants.

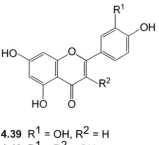

ME [C≡C]₃ CH=CH. CH=CH [CH₂]₄ CH=CH₂

4.38

4.39 R¹ = OH, R² = H
4.40 R¹ = R² = OH
4.41 R¹ = H, R² = OH

4.11 GLOBE ARTICHOKE

The globe artichoke, sometimes known as the cardoon, (*Cynaria cardunculus* or *C. scolymus*) is another member of the Asteraceae family. It is a perennial thistle. The fleshy lower portion of the flower bud is edible. Although it originates from the Mediterranean, it has been grown in Britain since the 16th century. The plant contains a high level of anti-oxidant phenolic compounds including cynarin (1,3-dicaffeoylquinic acid) **4.42** and chlorogenic acid (5-caffeoylquinic acid) and the flavone, luteolin **4.39** 7-glucoside. Extracts of the leaves of the artichoke have been found to have an antihyperlipidemic effect and to suppress serum triglyceride levels in test animals. The sesquiterpenes cynaropicrin **4.43**, aguerin B **4.44** and grosheimin **4.45** were isolated as the active components. They were accompanied by the related C-8 glycosides, the cynarascolides A–C. which inhibit hepatic cholesterol biosynthesis.

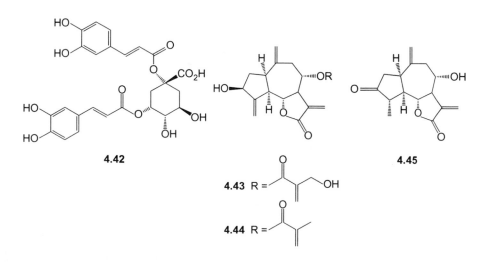

4.42

4.43 R =

4.44 R =

4.45

The essential oil from the plant contains a number of volatile compounds which contribute to the aroma including the sequiterpene hydrocarbons β-selinene (32%) **4.46**, β-elemene (5%) **4.47** and α-cedrene **4.48**, together with phenylacetaldehyde (12.9%) and benzyl alcohol (27%). The sesquiterpene β-elemene has attracted interest as an anti-tumour agent because it induces apoptosis and limits cellular differentiation. A series of triterpenoid saponins with the pentacyclic ursane skeleton, the cynarasaponins A–G, have also been isolated from globe artichoke. It is worth noting that the Jerusalem artichoke is a different plant (*Helianthus tuberosus*).

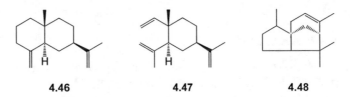

4.46   **4.47**   **4.48**

In this chapter, we have seen how the natural products that are characteristic of particular families, such as the glucosinolates of the Brassicaceae and the sesquiterpenoids of the Asteraceae, confer specific properties on these crops.

CHAPTER 5

Seed Vegetables

5.1 INTRODUCTION

The seeds of a number of plants provide valuable foodstuffs. Since they contain the enzymatic material to facilitate the growth of a new plant, as well as starch reserves, their protein content is often higher than that of other parts of the plant. Many seeds are contained within a protective coating with a structure and chemical content that can impede herbivorous predators. The legumes, of which the peas and bean are major examples, are widely grown as vegetables. The seeds are borne in pods which may also form an edible part of the crop when cooked. Peas and beans are members of the Fabaceae (Leguminoseae) family, which also includes a number of other plants that are found in the garden such as the lupin. These contain toxic alkaloids. Other seed crops, such as sweet corn, belong to different families. Sprouting seeds that are harvested within a few days of germination can be used to add flavour to salads and cooked dishes. The sprouts retain many of the beneficial nutrients of the seed and also include as flavouring components, compounds that have a role as protective agents for the developing seedling. Seeds that are used include the mung bean (*Vigna radiata*), azuki bean (*V. angularis*), black-eyed pea (*V. unguiculata*), chick pea (*Cicer arietinum*) and snow pea (mangetout pea) (*Pisum sativum* var. *saccharatum*) which belong to the *Fabaceae*, as well as some drawn from other families such as mustard and cress (*Sinapis alba* and *Lepidium sativum*), mizuna (*Brassica rapa*) and radish (*Raphanus sativus*).

Chemistry in the Kitchen Garden
By James R. Hanson
© James R. Hanson 2011
Published by the Royal Society of Chemistry, www.rsc.org

5.2 THE LEGUMES

A characteristic feature of the peas and beans is the symbiotic relationship that they have with *Rhizobium* species of nitrogen-fixing bacteria which are found in root nodules on these plants. Biological nitrogen fixation is mediated by an enzyme, nitrogenase, which reduces dinitrogen to ammonia. Nitrogenase contains two units. The smaller unit has a molecular weight of approximately 60 000 and contains four iron atoms. The larger unit is a complex tetrameric molybdenum–iron protein and has a molecular weight of approximately 220 000. It contains as many as thirty iron atoms and crucially two molybdenum atoms or, less commonly, two vanadium atoms. The larger of the two proteins carries out the actual reduction, whilst the smaller unit to which it is coupled is involved in the electron-transfer mechanisms. The ammonia which is formed is assimilated by the plant *via* glutamine **5.1**, the 5-amide of glutamic acid.

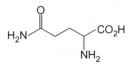

5.1

The biological fixation of nitrogen not only facilitates the synthesis of nitrogen-containing natural products by the plants but also means that these plants can be used as part of a crop rotation to replenish nitrogen levels in the soil. These plants have a higher protein content than many others, and hence a number, such as alfalfa, clover and vetch, are used as forage crops. Clover, vetch and a forage pea can be used as a 'green manure'. These plants can be sown in September after a vegetable plot has been used and, once they have grown, they can be dug in during the winter or early spring to improve the nitrogen content and organic material within the soil.

Although legume seeds are a rich source of dietary proteins, they also contain allergenic proteins. These have been identified not just in peanuts and soybeans but also in lentils, peas and beans. Due to the serious nature of these allergic reactions, considerable effort has been expended, particularly with soybean and peanuts, aimed at developing hypoallergenic cultivars which do not produce these proteins. Four protein families account for the majority of the legume allergens. These include storage proteins (cupins and prolamins), profilins and a group of pathogenesis related proteins.

The seeds of the legumes are also sources of dietary starch and fibre. One of the less pleasant aspects of eating peas and beans can be flatulence. The problem arises because the human intestinal mucosa lacks the enzyme system, α-galactosidase. As peas mature, the galactose derivatives of sucrose, known as raffinose and stachyose, accumulate (1.2 and 3.2 g per 100 g, respectively compared to sucrose, 6.2 g per 100 g). These oligosaccharides are not cleaved until they reach the lower intestinal tract where they are subject to bacterial metabolism by *Clostridium perfringens* and *Escherichia coli* to release carbon dioxide, hydrogen and methane.

5.3 GIBBERELLIN PLANT HORMONES IN THE LEGUMES

The gibberellin plant hormones (see section 1.13) play an important role in the maturation of seeds, in stem elongation and in leaf and fruit development. In the seed, they induce the synthesis in the aleurone cells, of the enzyme α-amylase which mediates the hydrolysis of starch. The size and availability of the seeds of the legumes have made them a useful subject for the study of the gibberellins in higher plants. Gibberellin A_1 **5.8** was the first of the gibberellin plant hormones to be isolated from a higher plant. It was found in 1958 in runner beans after extensive chromatography. The subsequent application of gas chromatography-mass spectrometry to the identification of gibberellins has enabled their widespread occurrence in higher plants to be established. There are now approximately 150 gibberellins that are known, all possessing the same underlying carbon skeleton but differing in their hydroxylation pattern. However, not all these gibberellins are biologically active. Many are intermediates on pathways leading to the active compound whilst others are catabolites formed from the active compound.

Unlike the gibberellins found in the Cucurbitaceae, the major hormones that are found in the Fabaceae (Leguminosae) contain a 13-hydroxyl group. The 13-desoxy series form minor components. However, within the Fabaceae there are chemotaxonomic differences. The garden pea (*Pisum sativum*) and the broad bean (*Vicia faba*), which belong to the tribe *Viciae*, produce gibberellins lacking a 3-hydroxyl group, whilst the beans (*Phaseolus coccineus* and *P. vulgaris*), which belong to the tribe *Phaseolus,* produce gibberellins with a 3-hydroxyl group. Even though the seeds of these plants are regarded as good sources of the hormones, the concentrations are typically at the level of μg per g fresh weight or even less.

The biosynthetic sequences which link these hormones at different stages of maturity of the seeds have been investigated. The gibberellins are diterpenoid substances which are formed by the oxidative metabolism of a hydrocarbon, *ent*-kaurene (see section 1.13). Gibberellin A_{12} aldehyde, which is the first gibberellin intermediate, is formed by the ring contraction of 7β-hydroxy-*ent*-kaurenoic acid. In plants, gibberellin A_{12} aldehyde is oxidized to the acid **5.2** and then in the legumes this is converted to a sequence of gibberellins with or without a 13-hydroxy group (**5.3–5.18**).

As the seeds approach maturity, the amounts of the biologically active gibberellins that are present decrease. In the garden pea, gibberellin A_{20} **5.7** is deactivated by hydroxylation at C-2 to give gibberellin A_{29} **5.12** which is then converted to the unsaturated ketone **5.19** as a catabolite. In the beans, gibberellin A_8 **5.11** is formed and converted to a glucoside conjugate. Enzyme systems have been obtained from peas and beans which mediate these hydroxylations. They require iron(II), 2-oxoglutarate, oxygen and ascorbate in order to function.

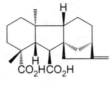

5.2

13-hydroxy series   13-desoxy series

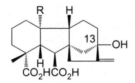   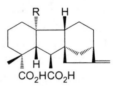

5.3 GA_{53} R = Me   **5.13** GA_{15} R = CH_2OH
5.4 GA_{44} R = CH_2OH **5.14** GA_{24} R = CHO
5.5 GA_{19} R = CHO
5.6 GA_{17} R = CO_2H

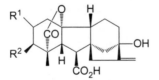   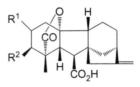

5.7 GA_{20} R^1= R^2= H   **5.15** GA_9 R^1= R^2= H
5.8 GA_1 R^1 = H, R^2= OH   **5.16** GA_4 R^1= H, R^2= OH
5.9 GA_5 R^1 = R^2 = ‖   **5.17** GA_{51} R^1= α-OH, R^2= H
5.10 GA_6 R^1 = R^2 = O   **5.18** GA_{34} R^1= α-OH, R^2= OH
5.11 GA_8 R^1 = α-OH, R^2= OH
5.12 GA_{29} R^1 = β-OH, R^2= OH

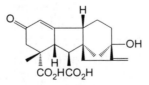

5.19

The sequiterpenoid plant hormones, abscisic acid **5.20** and phaseic acid **5.21**, have also been detected in the course of these studies.

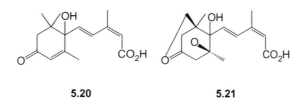

| 5.20 | 5.21 |

5.4 APHIDS ON THE LEGUMES

Plants of the pea and bean family are particularly sensitive to attack by aphids. Problems arise not just from the weakening of the plant from damage to the phloem but also because the aphid can act as a vector for plant viruses. Due to their parthenogenetic reproduction, aphids have short generation times such that a small initial population can rapidly lead to a significant infestation. The black bean aphid (*Aphis fabae*) over-winters on the spindle tree (*Eunonymus europaeus*) and in the spring it migrates to its summer host plants attracted by volatile plant metabolites. The volatile chemicals used in host location by the black bean aphid have been identified. These include common plant volatiles such as Z-3-hexen-1-ol and its acetate ester, hexan-1-ol, E-hex-2-en-1-al, benzaldehyde, 6-methylhept-5-en-2-one, (*R*)-linalool **5.22**, (*E*)-β-farnesene **5.23** and caryophyllene **5.24**. The ratio of these compounds in the mixture is important. (*E*)-β-Farnesene is also an alarm pheromone which is emitted by the aphids. (*Z*)-3-Hexen-1-ol and (*E*)-β-farnesene act not only as attractants to the predatory hoverfly, *Episyrphus balteatus,* but also induce oviposition. The essential oil of a South African wild mountain sage, *Hemizygia petiolata,* contains high levels of (*E*)-β-farnesene. This oil has been examined as a potential repellant for aphids such as the pea aphid, *Acyrthosiphon pisum.* In field plot experiments when the plants were

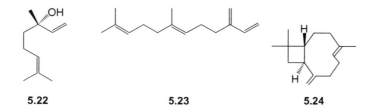

| 5.22 | 5.23 | 5.24 |

treated with a slow release formulation of the oil, there was a significant reduction in the numbers of *A. pisum*. After landing on a plant, the aphid inserts its stylet into the plant and senses the natural products within the plant to identify probing stimulants which enable it to discriminate between host and non-host plants. Some quercetin **5.25** glycosides have been identified as probing stimulants for a bean aphid, *Megoura crassicauda*. (*E*)-2-Methylbut-2-ene-1,4-diol 4-O-β-D-glucopyranoside **5.27** has been identified in a vetch, *Vicia hirsuta* which decreases the probing activity of the bean aphid.

The dark coloured aphids, *Aphis fabae,* which are found on broad beans, produce a pigment when they are damaged by predators such as ladybirds. This pigment readily stains clothing. The blood of these insects contains a glucoside, protoaphin **5.28**, which readily loses glucose to form a yellow unstable substance, xanthoaphin, which is converted to an orange pigment, chrysoaphin, and finally to a red pigment, erythroaphin **5.29**. This series of changes has been formulated as the cyclization and stepwise dehydration of **5.28** to give the perylene derivative **5.29**.

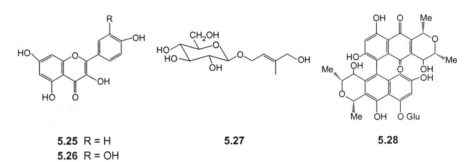

5.25 R = H
5.26 R = OH
                                                5.27   **5.28**

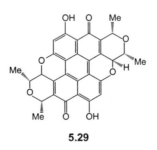

5.29

Aphids feed on the sap of a plant and exude excess sucrose as a solution, honeydew. This acts as an attractant for ants which in turn appear to protect the aphids. However, the sticky surface of the plant then becomes a good culture medium for microbial attack. The ladybird

(*Coccinella septempunctata*) feeds on aphids. It utilizes 2-isopropyl-3-methoxypyrazine as an aggregation pheromone. The ladybird also produces a number of polycyclic quinolizine alkaloids such as coccinelline as feeding deterrents for birds.

5.5 PEAS

Not only do peas (*Pisum sativum*) have a higher protein content than many other vegetables, but they also synthesize some non-proteogenic amino acids which may play a role in defense mechanisms against herbivores. Homoserine together with two uracilylalanines, willardiine **5.30** and isowillardiine **5.31** were isolated from growing pea shoots. Willardiine has attracted interest because of its action on the central nervous system.

Another non-protein amino acid, β-(3-isoxazolin-5-one-2-yl)-L-alanine **5.32**, has been found in the seedlings and root exudates of peas. This amino acid has growth inhibitory activity towards the seedlings of other plants such as lettuces. This allelopathic role, whilst beneficial to the pea, may limit the other crops that might be grown alongside peas. This amino acid is unstable and breaks down to give α,β-L-diaminopropionic acid. This is a precursor of the neurotoxic amino acid, β-*N*-oxalyl-α,β-L-diaminopropionic acid. This amino acid which occurs in some other members of the Fabaceae (*e.g.* the grass pea, *Lathyrus sativus*) can give rise to the crippling effects of lathyrism.

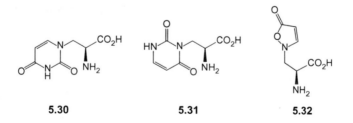

5.30   **5.31**   **5.32**

The seeds of the parasitic plants belonging to the *Striga* and *Oro-banche* genus germinate when they are stimulated by exudates from potential host plant roots. In some parts of the world, these parasitic plants present a serious problem with peas and beans. The strigo-lactones, *e.g.* strigol **5.33**, have been identified as germination stimulants. A relative, fabacyl acetate **5.34**, has been detected as a major component of the root exudate of peas and beans. Peagol **5.35** and

peagoldione **5.36** are two further compounds possessing the characteristic hydroxybutenolide ring, which have been isolated from pea root
exudates.

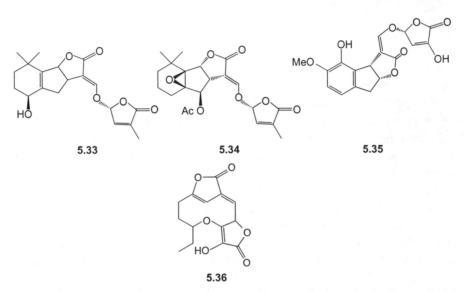

5.33   **5.34**   **5.35**

5.36

Isoflavones and their relatives are characteristic metabolites of the
legumes. In many cases they are produced under stress conditions. An
important subset of the isoflavones is the pterocarpans. These compounds occur as phytoalexins. Although the concept of a phytoalexin
was proposed in 1940, the first phytoalexin to be isolated was pisatin
which was obtained from peas in 1960. The structure of pisatin **5.37**
was established in 1962. It was obtained from the endocarp of peas
that had been infected with the fungus, *Monolinia fruticola,* and it
shows antifungal activity at a concentration of 10^{-4} M. The biosynthesis occurs from an isoflavone *via* sophorol **5.39**. Pisatin is less
toxic to a pea pathogen, *Ascochyta pisi,* which owes its virulence to
its ability to metabolize pisatin to the less active (+)-6a-hydro-
xymaackain **5.38** by dealkylation of the methoxyl group. Some isoflavonoids have a formal similarity to the steroid hormone, estradiol,
and are known as phytoestrogens. This structural similarity leads to
perturbations of estrogenic effects through their ability to bind to the
estrogen receptor. The presence of phenolic hydroxyl groups and the
distance between them are important factors affecting this activity.
Apart from peas and beans, they are also found in soybeans and in
flax.

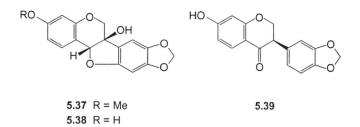

5.37 R = Me
5.38 R = H

5.39

The widespread pentacyclic triterpenes, α- and β-amyrin, have been found in peas and their biosynthesis has been studied in this plant. The glycosides of more highly hydroxylated derivatives, isolated from peas, are known as the 'soyasaponins' because they have also been isolated from soybeans. They have shown insecticidal and insect antifeedant activity against insects that are pests of stored food products. They were isolated alongside some lysolecithins which have a detergent-like activity and which enhanced their biological activity.

2-Alkyl-3-methoxypyrazine together with 2,6-nonadienal and 2,4-decadienal are responsible for part of the odour of green peas.

5.6 BROAD BEANS

The broad bean (*Vicia faba*) had its origin in North Africa and was quite widely cultivated around the Mediterranean before being brought to Northern Europe. Amongst the amino acids and their relatives produced by broad beans are L-tyrosine, tyramine and L-dihydroxyphenylalanine (L-DOPA) **5.40**. The presence of L-DOPA was first established in 1913. More recently, the broad bean was considered as a source of this amino acid which plays an important part in the treatment of Parkinson's disease. Jasmonyl conjugates of these amino acids have been isolated from the flowers of the broad bean. The major gibberellins which were identified in *V. faba* were gibberellins A_{17} **5.6**, A_{19} **5.5**, A_{20} **5.7**, A_{29} **5.12**, A_{44} **5.3** and A_{53} **5.4** representing a sequence hydroxylated at C-13 but not at C-3, unlike those found in *Phaseolus*. Vicine **5.41** is a constituent of broad beans and some vetches. It is the 5-β-D-glucoside of a pyrimidine, divicine. This aglycone is released in the intestinal tract by β-glycosidases and in some susceptible people it can cause a haemolytic anemia known as favism. When broad bean leaves were infected with the fungal pathogen, *Botrytis cinerea*, they responded by producing wyerone acid **5.42** in amounts ranging from 3–30 μg per g fresh tissue. This phytoalexin was named after Wye College where it was first isolated.

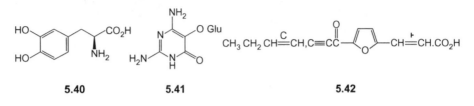

5.40   **5.41**   **5.42**

5.7   RUNNER BEANS

The runner bean (*Phaseolus coccineus*) and the French bean (*Phaseolus vulgaris*) are very useful garden vegetables. The compounds which they contain have many similarities to those found in peas although there are differences which have been noted in their gibberellin content.

Runner beans produce a range of isoflavonoids which increase when the plant is subject to stress. Several of these stress metabolites are fungitoxic. The major components include genistein (1 µg per g fresh weight) **5.43**, 2′-hydroxygenistein (0.6 µg per g) **5.44**, kievitone (1.1 µg per g) **5.45**, and phaseolin (1.7 µg per g) **5.46**. The latter has been studied as a phytoalexin. These beans also contain flavonol glycosides based on kaempferol **5.25** and quercetin **5.26** in which the sugar units are attached to C-3. Beans provide a dietary source of these flavonols. It has been suggested that one of the quercetin derivatives may act as an oviposition deterrent for butterflies. The red colour of some beans arises from glycosides of pelargonidin **5.47**. The leaves of *Phaseolus vulgaris* contain the amide conjugate salicyloyl aspartate derived from salicylic acid and aspartic acid.

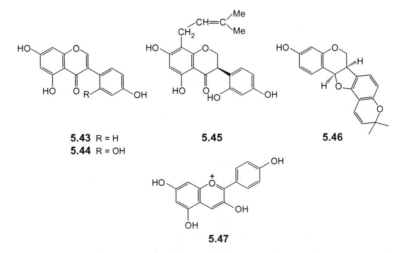

5.43 R = H   **5.45**   **5.46**
5.44 R = OH

5.47

5.8 SWEET CORN

Maize (*Zea mays*) is a member of the Poaceae family which consists of grasses and, like wheat and barley, it is grown as a large scale field crop. Sweet corn (*Zea mays var. saccharata*) has, as its name suggests, a higher sugar content than normal maize and some varieties can be grown as a vegetable in the kitchen garden. It is a monoecious plant, with male tassels and the female ears which generate the cobs, appearing on the same plant. The female flowers are wind pollinated and hence good pollination is achieved by growing sweet corn in blocks rather than rows. Sweet corn is harvested at a stage when the cob is still milky, before the kernals are dry and before all the sugar is converted to starch. The formation of starch continues after harvest and can lead to a loss of taste. Fresh sweet corn does not store well although it can be frozen. Tinned sweet corn has been 'blanched' during processing to destroy the enzymes. The presence of soluble sugars and a wide range of amino acids in corn have made corn steep liquor a valuable growth medium for fungi. One of the major early improvements to the penicillin fermentation in the war years involved the use of corn steep liquor in the fermentation broth. This is also underlines a warning for harvesting sweet corn which can be easily attacked by fungi. In order to soften maize and remove the husks before making cornflour, it was soaked in water which was often made alkaline. This led to a problem because niacin (nicotinic acid) was removed, causing the deficiency disease, pellagra.

Apart from its carbohydrate, amino acid and vitamin content, sweet corn contains carotenoids which impart a yellow colour. The carotenoids which have been detected include zeaxanthin **5.48** and its double bond isomer, lutein **5.49**, together with lesser amounts of α- and β-carotene and cryptoxanthin. Zeaxanthin has been found in other foodstuffs such as broccoli and spinach. A high dietary intake of foods containing zeaxanthin has been associated with a lower incidence of age-related macular degeneration of the eyes. Cyanidin 3-glucoside is the major anthocyanin that is present in red popcorn. Pelargonidin **5.47** and peonidin glucosides have also been found in maize.

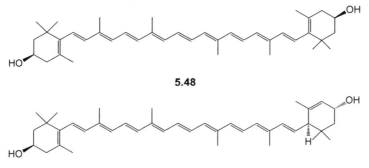

5.48

5.49

The tocopherol antioxidants are also present. Cooking sweet corn can lead to the extraction of the phenolic acids, *p*-coumaric acid, ferulic acid, 5-hydroxyferulic acid and sinapic acid from the cell walls. The plant sterols that have been detected include sitosterol (24-ethylcholest-5-en-3β-ol) and campesterol (24-methylcholesterol) which have a beneficial inhibitory effect on cholesterol deposition in the body.

The gibberellin plant hormones which have been identified in maize belong to the 'early' C-13 hydroxylation pathway and include the sequence GA_{53} **5.3**, GA_{44} **5.4**, GA_{19} **5.5**, GA_{20} **5.7**, GA_1 **5.8** and GA_8 **5.11**. Other members of this series which have been found in maize include GA_{17} **5.6** and GA_{29} **5.12**. Gibberellin A_1 **5.8** is the active metabolite inducing stem elongation. Dwarf maize mutants have been identified in which key steps in the gibberellin biosynthetic pathway are blocked.

The leaves, sheath and husks of maize produce some volatile sesquiterpenes including bisabolene, sesquithujene **5.50** and curcumene **5.51** hydrocarbons. Those which appear after lepidopteran herbivore attack, attract parasitic wasps which use the larvae as their hosts. The epicuticular wax on the leaves which comprises long-chain n-alkane hydrocarbons, such as n-hexacosane ($C_{26}H_{54}$) and n-heptacosane ($C_{27}H_{56}$), acts as an oviposition stimulant for the European corn borer, *Ostrinia nubilalis*.

Many graminaceous plants such as wheat and maize produce cyclic arylhydroxamic acids as defensive agents. The major benzoxazinoids in maize are 2,4-dihydroxy-7-methoxy-(2H)-1,4-benzoxazin-3(4H)-one (DIMBOA) **5.52** and its demethoxy derivative, DIBOA. These are stored in the plant as the 2β-O-D-glucosides and released when the plant is damaged. These arylhydroxamic acids are toxic to a wide range of organisms including the European corn borer, various aphids and some plant pathogenic fungi. They are biosynthesized from indole by insertion of oxygen into 3-hydroxyindolin-2-one to form 2-hydroxybenzoxazinone. *N*-Hydroxylation and oxidation of the benzene ring then follows. In wheat, it has been shown that the

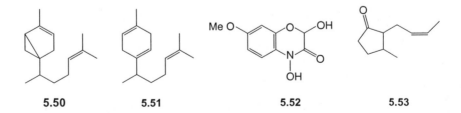

5.50   5.51   5.52   5.53

volatile activator, *cis*-jasmone **5.53** which is produced from jasmonic acid, induces the accumulation of DIMBOA, as well as some plant defensive phenolic acids such as ferulic acid. Although not available in the UK, genetically modified maize has been used in the USA to combat herbivorous insects and to control weeds. One form of genetically modified maize incorporates the gene that codes for the formation of the *Bacillus thuringiensis* (Bt) insect toxin, whilst another is resistant to the glyphosate herbicide, Roundup®.

CHAPTER 6
Greenhouse Crops

6.1 INTRODUCTION

A small greenhouse or a cold frame provides a controlled environment in which various garden crops can flourish. These form the subject of this chapter. In many cases, these plants had their origin in the sub-tropical regions of the world where there is strong sunlight, and significant insect and mammalian predation. It is not surprising, therefore, that they produce pigments such as the carotenoids with intense light-absorbing properties which can act as photoprotective agents and, in association with chlorophyll, act as accessory light-harvesting pigments. In addition, some of their hot or bitter-tasting constituents, more commonly present in 'wild' strains, have insect deterrent activities. However, it is worth noting that in some parts of the UK particular cultivars of these plants can flourish out-of-doors.

6.2 TOMATOES

Tomatoes (*Lycopersicon esculentum,* sometimes known as *Solanum lycopersicum*) are members of the Solanaceae family which also includes the potato. Consequently there are some natural products which are common to both. The tomato originated from Central and South America and it is reported to have been brought back to Europe by

Chemistry in the Kitchen Garden
By James R. Hanson
© James R. Hanson 2011
Published by the Royal Society of Chemistry, www.rsc.org

Spanish explorers in the 16th century. However, it had probably been domesticated centuries earlier by the Aztecs. There are many cultivars of the tomato with fruit which range in size from small 'cherry' tomatoes, to 'plum' tomatoes and large 'beef-steak' tomatoes. Although the characteristic fruit is red, there are yellow, purple and very dark cultivars. Some cultivars have a bushy determinate form and tend to bear all their fruit at the same time, whilst other indeterminate varieties develop into vines and produce their fruit over a period of time. Finally, some varieties are best grown in the greenhouse whilst others will grow outdoors in temperate climates. Consequently, with this range of cultivars there is some variation in their constituents.

Like the potato, the tomato produces steroidal alkaloids. The alkaloid, α-tomatine **6.1**, consists of an aglycone (tomatidine) with a tetrasaccharide (β-lycotetraose) attached to the 3-position. The sugar comprises two molecules of glucose and one each of galactose and xylose. This toxic alkaloid is found in the leaves, stems, roots and flowers, and to a small extent in the green fruit of the tomato. It provides protection against herbivores and microbial attack. The fungus, *Fusarium oxysporum,* is a widespread soil-borne plant pathogen which causes general vascular wilts. A relative of this fungus, *F. lycopersici,* which is a successful pathogen on tomatoes, can deactivate α-tomatine by cleaving the sugar to give the inactive components, tomatidine and the sugar. The grey powdery mould, *Botrytis cinerea,* can also deactivate α-tomatine in this manner. The tomato pathogen, *Alternaria solani,* removes the sugars to give a monosaccharide.

There are other steroidal alkaloid glycosides known as the lycoperosides **6.2** which have been isolated from tomato fruit. These contain additional oxygen functions on the heterocyclic ring F. In over-ripe tomato fruit when their protective effect is presumably no longer required, these alkaloids appear to undergo degradation to pregnane glycosides, for example 3-O-β-lycotetraoxyl-3β-hydroxy-5α-pregn-16-en-20-one **6.3**. At one time the steroidal alkaloids obtained from tomato plants, particularly the seeds, were considered as possible raw materials for the preparation of the hormonal steroids required by the pharmaceutical industry.

Tomato leaves and fruit also contain tryptophan derivatives which may also act as precursors of the auxin, indolylacetic acid. Under conditions of water stress, tryptophan is converted to the *N*-malonyl derivative. There are some reports that the D-enantiomer of tryptophan may be involved as an indolylacetic acid precursor. Another tryptophan metabolite which has been detected is the formaldehyde cyclization product, lycoperodine-1 **6.4**.

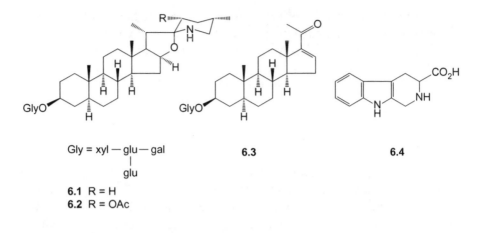

Gly = xyl — glu — gal
            |
        glu

6.1 R = H
6.2 R = OAc

6.3

6.4

A number of gibberellin plant hormones have been identified in the foliage of the tomato. As we have seen in previous chapters, the gibberellin plant hormones can be grouped into biosynthetic sequences based on their hydroxylation pattern. Sequences can be differentiated depending on whether a hydroxyl group is inserted at C-3 or C-13 immediately after ring contraction, or at a much later stage in the biosynthesis when the C_{19} compounds with the γ-lactone ring have been formed. Most members of an early 13-hydroxylation series have been identified in the foliage of the tomato starting with GA_{53} and proceeding *via* GA_{44}, GA_{19}, GA_{20} to GA_1 and GA_{29} (for structures see Chapter 5, **5.3–5.12**). A minor early non-3,13-hydroxylation pathway also appears to exist. The results of experiments involving the application of various gibberellins and inhibitors of gibberellin biosynthesis have shown that the setting of tomato fruit after pollination is gibberellin dependent. Gibberellin A_1 was the most active gibberellin for tomato fruit growth. These studies showed that pollination increased the gibberellin levels in the ovary by increasing the expression of the genes encoding the gibberellin C-20 oxidase. There was no evidence for any effect of pollination on gibberellin inactivation through conversion to gibberellin A_8.

The ripening of the fruit of the tomato takes place in a co-ordinated way in which ethylene gas plays an important hormonal role. Tomatoes are a climacteric fruit which have a high respiration rate during ripening which can continue after they have been picked. This sensitivity to ethylene explains the method of ripening green tomatoes in the autumn by picking them and putting them under newspaper with a ripe tomato or a banana as the source of the ethylene. Ethylene, which is also formed by a number of other fruits such as apples and pears, is biosynthesized from 1-aminocyclopropane carboxylic acid. The post-harvest ripening of fruit is delayed commercially by storing the fruit in the presence of 1-methylcyclopropene which binds to the ethylene receptor.

The most obvious change as the tomato ripens is in the carotenoid pigments. There is a progressive desaturation from the hydrocarbon phytoene **6.5** and its 15(Z)-isomer, phytofluene, to ξ-carotene and the more highly conjugated red pigment, lycopene **6.6**. In the ripe tomato, this pigment is present in the range 9–40 μg per g fresh weight. Although lycopene is not a vitamin A precursor, it has an important biological activity as a powerful antioxidant. A more readily absorbed form of lycopene, Ateronon®, has been examined in the context of lowering cholesterol levels. A number of aldehydes, the *apo*-lycopenals, *e.g.* **6.7**, are also formed as a result of oxidative degradation of this carotenoid. Phytoene 1,2-oxide is also formed. The colouring matter of purple tomatoes includes some anthocyanins. Flavonoid conjugates derived from quercetin have also been found in tomatoes.

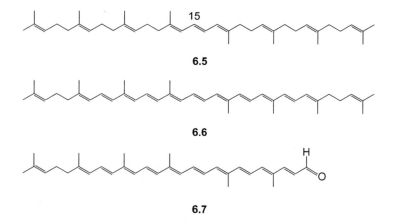

Apart from colour changes, ripening is accompanied by changes to the aroma, flavour and texture of the tomato. Over 400 volatile compounds have been detected from various tomato cultivars. Their contribution to the aroma of the tomato has been assessed by considering an odour unit. The log. odour unit (log U) is defined as the logarithm of the ratio of the concentration of the volatile compound to its odour threshold. Compounds with a positive log. odour unit are assumed to make a significant contribution to the flavour, whilst those with a negative value may only be making a background contribution. The compounds which contribute to the odour of the tomato fall into three groups. There are those which are formed and remain as volatile compounds. Then there are those which are formed and then stored as their non-volatile glycosides, and finally there are those which are formed by an oxylipin or oxycarotene pathway as the tomato ripens. The major tomato volatiles include (Z)-3-hexenal, hexanal, l-penten-3-one, (E)-2-hexenal, (E)-2-heptenal and (Z)-3-hexenol which are probably formed by an oxylipin pathway. Esters derived from these were also found. The 2- and 3-methylbutanols, 2-isobutylcyanide and 2-isobutylthiazole are formed from the amino acids, leucine and isoleucine. Some terpenoids including limonene, α-pinene and linalool, as well as carotenoid degradation products such as β-ionone **6.8**, together with geranyl **6.9** and farnesylacetone are also present. Some of these are stored as their glycosides. Compounds from the phenylpropanoid pathway include eugenol **6.10**, phenethyl alcohol, benzaldehyde and methyl salicylate. Although the concentration of the major volatiles increases as ripening progresses, the production of methyl salicylate decreases. The latter has a role as a marker of infection. The volatiles that are released from infected leaves differ from those that are produced by tomato fruit. Tomatoes grown outdoors have been shown to be richer in volatile constituents compared with those grown in the greenhouse or artificially ripened from green fruit.

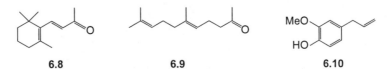

6.8   **6.9**   **6.10**

As ripening progresses the amount of starch that is present decreases and the concentration of sugar increases. The amylose in the starch decreases more rapidly than the amylopectin. The total acidity also decreases. However, whilst some amino acids such as alanine decrease, there is a marked increase in the amount of glutamic acid which is present. This can reach as much as 200 mg per 100 g fresh weight. Sodium glutamate is well-known as a flavour enhancer and it is possible that this, together with the colour, may account for the use of tomato as an additive to many dishes such as pizzas. Another effect of the ripening hormone, ethylene, is to increase the activity of the N-glycans and the degradation of the N-glycoprotein constituents of the cell walls leading to their increased flexibility and the tendency of over-ripe tomatoes to split.

6.3 AUBERGINES

The aubergine or egg plant (*Solanum melongena*) is a relative of the tomato, which can be grown as a greenhouse crop. However, the edible fruit are normally cooked. The seeds within the fruit contain steroidal alkaloids such as solasodine, a small amount of nicotine and some steroidal saponins known as the melongosides **6.11**. The furostanol glycosides have sugar units attached to C-3 and to C-26. The fruit also possess a high antioxidant activity associated with the presence of chlorogenic acid and a number of its isomers, together with anthocyanins. The major purple pigments in the aubergine skin are the anthocyanins, nasunin [delphinidin **6.12** 3-(*p*-coumaroylrutinoside)-5-glucoside] and tulipanin (delphinidin-3-rutinoside). These anthocyanins, particularly nasunin, have a high antioxidant radical scavenging ability.

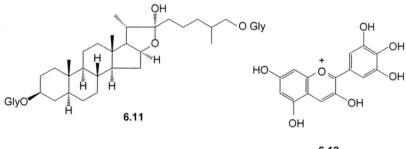

6.11

6.12

6.4 PEPPERS

Peppers (*Capsicum annuum*) are another widely grown greenhouse fruit. There are many cultivars differing not only in their colour and shape but also in their use as a spice. There are cultivars that are green, yellow or red, and some that are mild tasting and others that are very 'hot' tasting.

The major pigments are carotenoids which change during ripening. The yellow fruit contain lutein (3,3'-dihydroxy-α-carotene) and the epoxide, violaxanthin. The red fruit contain β-carotene and the rearrangement products, capsanthin **6.13** and capsorubin **6.14**. In green peppers which retain their chlorophyll, the carotenoid pigment which is present is lutein. As the yellow or red peppers ripen and the chlorophyll

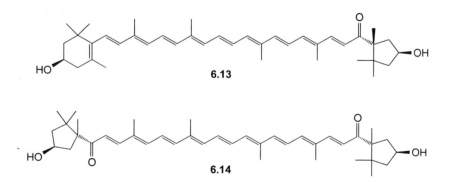

6.13

6.14

disappears, the carotenoid content increases substantially. The epoxides such as violaxanthin undergo rearrangement to form capsanthin and capsorubin and these in turn undergo epoxidation and oxidative cleavage to form apocarotenoids. Capsanthin and capsarubin have been authorized for use in the EU to colour foods and cosmetics.

The hot cultivars contain the capsaicinoids. These alkaloids are acid amides of vanillylamine and have potent neurological effects. The best known of these is capsaicin (8-methyl-*N*-vanillyl-6-nonenamide) **6.15** which also occurs as its glucopyranoside. There is a specific capsaicin receptor or vanilloid receptor. Capsaicin creams are beneficial to patients with psoriasis. Some of the sweet non-pungent peppers contain non-nitrogenous analogues of capsaicin based on vanillyl alcohol.

The volatile components of peppers include (*E*)-hept-3-en-2-one and (*E*)-non-2-en-4-one, 3-isobutyl-2-methoxypyrazine, linalool and β-damascenone **6.16**, as well as a dithiolane **6.17**. When peppers undergo attack by fungi they produce a sesquiterpenoid antifungal phytoalexin, capsidiol **6.18**.

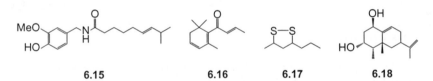

| 6.15 | 6.16 | 6.17 | 6.18 |

6.5 THE CUCURBITS

Cucumbers, courgettes, marrow, squash and pumpkins are all members of the Cucurbitaceae family. These plants are spreading vines, many of which originated in the tropics. Cucumbers are reported to have come from India. Although they are often grown outdoors particularly where the soil is warm as on a compost heap, some cucurbits can benefit from being grown in an enclosed environment such as a cold frame, despite the fact that they can occupy quite a lot of space and not all cultivars are self-fertile. Pumpkins require the warmth of summer and plenty of water.

Cucurbits suffer from a number of fungal diseases which attack the leaves and stems, as well as the developing fruit. The composition of the volatile material from the flowers has been examined in order to identify attractants for the cucumber beetles. These pests not only damage the leaves but also spread diseases. Indole and simple aromatic compounds including veratrole (1,2-dimethoxybenzene) and phenylacetaldehyde have been identified in this context.

6.6 CUCUMBERS

The edible portion of the cucumber (*Cucumis sativus*) is the fruit, although it is usually consumed as a vegetable. The cylindrical form of the cucumber is eaten in the unripe green state. The ripened yellow cucumbers are quite bitter. The cucumber contains a high percentage of water. Cucumber flavour is associated with the presence of (*E,Z*)-2,6-nonadienal, together with (*E*)- and (*Z*)-2-nonenal, 2- and 3-hexenals and (*Z*)-1,5-octadien-3-one. These compounds are derived by the oxylipin degradation of unsaturated fatty acids such as linolenic acid [(all-*Z*)-9,12,15-octadecatrienoic acid], linoleic acid [(*Z,Z*)-9,12-octadecadienoic acid] and oleic acid. The bitter taste of some cucumbers arises from the presence of cucurbitacin C **6.19**. The highly oxidized triterpenoid cucurbitacins are poisonous substances having diuretic and laxative action, as well as reducing blood pressure and showing antirheumatic activity. Extracts of the seeds were used in folk medicine as purgatives, whilst a cucumber cream has been used as a cosmetic. The cucurbitacins act as defensive substances against insect attack, for example against the spider mite, (*Tetranychus urticae*). The spider mite is a common greenhouse pest which, by feeding on leaves, removes the chlorophyll and reduces photosynthesis. When they are damaged, the leaves emit herbivore-induced plant volatiles, such as β-ocimene and farnesene that attract a predatory mite, *Phytoseiulus persimilis*, which attacks the spider mite.

The cucurbitacins have also been studied in pumpkins. The biosynthesis of cucurbitacin C **6.19** from squalene-2,3-epoxide *via* 10α-cucurbita-5,24-dien-3β-ol **6.20** has been studied in microsomal preparations from pumpkin (*C. maxima*) seedlings. The seeds also contain an unusual amino acid, β-pyrazol-1-ylalanine **6.21**, which is a histidine analogue. It is formed from pyrazole and serine and is also found as a γ-glutamyl dipeptide. It is possible that this amino acid may also contribute to the biological activity of cucumber preparations by mimicking histidine.

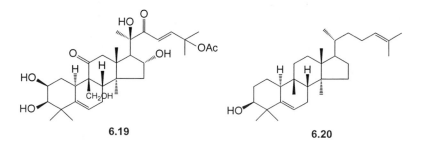

6.19   **6.20**

When cucumber leaves are infected by a fungus, *Sphaerotheca fuliginea* which produces a powdery mildew, they form some acylated flavone C-glycosides based on isovitexin (apigenin-6-C-glucoside) **6.22** as phytoalexins. The production of these phytoalexins is stimulated by treatment with potassium silicate.

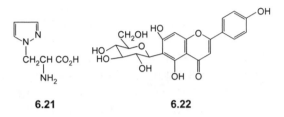

6.21   **6.22**

6.7   PUMPKINS, MARROWS AND COURGETTES

Pumpkins, marrows and courgettes (zucchini) (*Cucurbita maxima, C. moschata* and *C. pepo*) are rich in carotenoids and because, once harvested, they can be stored at room temperature for some months, they have a useful dietary role. The major carotenoid is the pro-vitamin A, β-carotene **6.23**, which may be found at concentrations as high as 30–60 µg per g in the mature fruits. α-Carotene, lutein **6.24**, violaxanthin **6.25** and

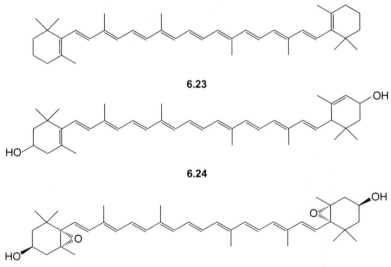

6.23

6.24

6.25

neoxanthin, together with a number of other oxygenated carotenoids have been detected. Lutein and violaxanthin also occur as esters of myristic and palmitic acids. Apart from β-carotene acting as a vitamin A precursor, other roles in human health have been attributed to these carotenoids in terms of their antioxidant activity and the reduction of the risk from degenerative diseases.

Vitamin E (α-tocopherol) is another powerful antioxidant which is found in many plant oils such as those of the pumpkin, where it serves to prevent autoxidation of unsaturated fatty acids by oxygen. One of its roles in plants and in humans may be to preserve membrane integrity from oxidative damage. The tocopherols, like chlorophyll, contain a C_{20} phytol chain. This is biosynthesized from geranylgeranyl diphosphate. In another pathway, this is dimerized to form phytoene and thence the carotenoids.

Pumpkin seeds and pumpkin seed oil are often consumed separately. The vitamin E content of pumpkin seeds can be quite high. The main vitamin E isomers, α- and γ-tocopherol have been detected at concentrations of up to 90 and 600 μg per g. The major fatty acids are palmitic (9.5–14.5%), stearic (3.1–7.4%), oleic (21.0–44.9%) and linoleic acids (35.6–60.8%). The predominance of unsaturated fatty acids is clear. Pumpkin seeds are also quite rich in plant sterols which are of interest because of their ability to lower serum cholesterol levels. Pumpkin seed oil is a constituent of health food medicines that are used to treat benign prostate hyperplasia. The pumpkin sterols, *e.g.* 24-ethylcholest-7-en-3β-ol are unusual in that they contain a Δ^{7}- rather than a Δ^{5}-double bond. Pentacyclic triterpenes of the multiflorane type esterified with *p*-aminobenzoic acid, a component of folic acid, have been found (0.02–0.08%) in pumpkin seed oil.

Beneficial polyphenols of the isoflavone type such as daidzein **6.26** and genistein **6.27**, and the lignan, secoisolariciresinol, have also been detected. Secoisolariciresinol diglucoside **6.31** is an antioxidant which can also reduce blood cholesterol levels. It is converted by intestinal bacteria into enterodiol and enterolactone **6.32** which may have beneficial effects on various disease states. The seeds of *Cucurbita pepo* also contain a group of phenolic glycosides with different acyl groups attached to the sugars. These are known as the cucurbitosides, *e.g.* **6.28**.

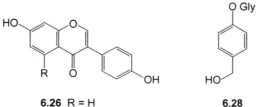

6.26 R = H
6.27 R = OH

6.28

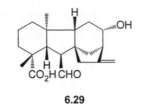

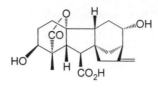

6.29

6.30

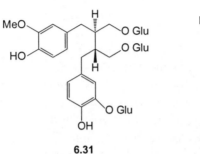

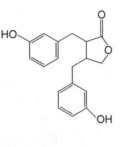

6.31

6.32

The availability of the pumpkin seeds has meant that they are potential substrates for studying various biosyntheses such as those of the carotenoids and the gibberellin plant hormones. A biosynthetic sequence of gibberellins from gibberellin A_{12} aldehyde to gibberellin A_4 and from 12α-hydroxygibberellin A_{12} aldehyde **6.29** to 12α-hydroxygibberellin A_4 (GA$_{58}$) **6.30** and its 2β-hydroxylation product (GA$_{49}$) have been identified in extracts of the endosperm and embryos from the seeds of *Cucurbita maxima*.

CHAPTER 7

Fruit Trees and Bushes

7.1 INTRODUCTION

The fruit trees and bushes which are grown in the garden provide both decorative and rewarding plants. Once established, they can provide crops with relatively little maintenance except for pruning. There was a time when fruit trees would inevitably occupy a large area, restrict the growth of other plants and, because of their height, present difficulties in harvesting. However, the development of dwarfing root-stocks has meant that apples, pears, plums and cherries can be grown in quite small gardens or in containers on a patio. It is even possible to graft several cultivars onto the same root-stock. The same is true of fruit bushes, where the ability to grow plants in containers has the added advantage that bushes, such as the blueberry which belong to the Ericaceae and require an acidic soil, can be grown almost anywhere. A number of hybrid fruits have been developed particularly by crossing different berries. Furthermore, small trees and bushes are also easier to protect from birds and from the effects of a late frost.

The phytochemicals produced by fruit, particularly the polyphenols, confer considerable health benefits. Although these have come to the fore in recent years, there are many descriptions in older herbals and books on folk medicine of the benefits associated with the consumption of fruits.

The fleshy edible part of fruits serves to protect the seeds whilst they mature. Once the seed has developed, the fruit ripens and assists the dispersal of the seed and its eventual germination. Consequently, fruit contain natural products which facilitate these roles. The anthocyanin,

Chemistry in the Kitchen Garden
By James R. Hanson
© James R. Hanson 2011
Published by the Royal Society of Chemistry, www.rsc.org

flavonol and carotenoid pigments found in fruit absorb light in the damaging ultra-violet region and they also behave as antioxidants to protect the seeds. There are carbohydrates which provide energy reserves and flavouring substances which can act as semiochemicals. There are many physiological changes which take place to fruit during ripening and these are reflected in changes to their components. The softening of fruit is primarily due to a change in the cell wall carbohydrate metabolism. This leads to a decrease in the neutral sugar residues in the polymers that are involved in the cell walls. There are changes to the acidity, the anthocyanin pigments, the plant hormones and the volatile organic compounds that are present.

In the early stages of the development of fruit, there are compounds which are present to deter insects, birds and animals but in the later stages these are replaced by compounds which act as attractants. Birds and mammalian herbivores may then be attracted by colour and taste to eat the fruit. Eventually they excrete the seed some distance from the plant in fertile manure to enhance the biodiversity of an area. Dung beetles and worms may even bury the seed and facilitate its successful germination and subsequent development of a young seedling. The lignin coating of the seed reduces its bacterial degradation. In this chapter, we will see the role of individual natural products in fruit and the development of the seeds.

7.2 APPLES

Apples are one of the most widely consumed fruit, and with the development of dwarf varieties one of the most commonly planted fruit trees in the garden. There is a widely held perception that apples are good for health, and this is supported by epidemiological studies which show an inverse correlation between the regular consumption of apples and the development of many chronic diseases of humans: 'An apple a day keeps the doctor away'. There are many different cultivars of 'eating' apples which differ in their aroma, flavour, acidity, astringency, sweetness, crispness or juiciness. Other cultivars are used as 'cooking' apples or for making cider. Not surprisingly there are wide variations in their constituents which may also change as the apple approaches maturity and during post-harvest storage or in the fermentation to make cider.

The main sugars are fructose and sucrose (approximately 4–5 g each per 100 g fresh weight) and glucose (approximately 1 g per 100 g) with smaller amounts of sorbitol. The major organic acid is L-malic acid (up to 1 g per 100 g) with smaller amounts of citric and succinic acids. The relatively small amounts of amino acids such as aspartic and glutamic acids (2–3 mg per 100g fresh weight), glycine, valine and leucine decrease as the apple matures.

Over two hundred compounds have been detected in the volatile products from apples. However, their contribution to our perception of the aroma of the apple depends not just on their individual concentration, but also on the threshold at which the human nose can detect them and on their combined effect with other substances. There are significant differences between the volatile constituents of different varieties. As the fruit matures and is harvested and stored, the profile of these compounds changes. Rainfall and temperature during the growing season have also been shown to influence the volatile compounds that are produced by the apples and the apple trees. In one study of the monoterpene emissions, linalool formed 94% of the monoterpene emissions at flowering, but only 11% at fruit ripening when it was accompanied by limonene (26%), *p*-cymene (21%) and camphor (17%). The major volatile compounds from the fruit include esters such as butyl acetate, 2-methylbutyl acetate and esters of (+)-2-methylbutanoic acid, ethyl and hexyl butanoate, hexyl acetate, hexenyl acetate, (*E*)-2-hexenal, (*E*)-2-hexenol and β-damascone **7.10**. Some apple cultivars accumulate (*R*)-(+)-octane-1,3-diol. As the apple matures, the concentrations of various sharp-smelling aldehydes such as butanal, pentanal, (*E*)-hex-2-en-al and heptanal declines, whilst that of the alcohols and esters increases. Some of the alcohols are stored as their glycosides and released by the action of glycosidases.

Many of the straight chain compounds arise from the oxylipin degradation of fatty acids, such as linoleic acid, whilst 2-methylbutanol and 3-methylbutanol arise from the isoleucine and leucine pathways of amino acid metabolism. Other compounds arise from the terpenoid and phenylpropanoid pathways. β-Damascone **7.10** is probably a carotenoid degradation product.

The presence of various polyhydroxylic phenols, such as chlorogenic acid, are not only associated with the antioxidant properties of the apple but also account for the browning of apples. As the apple is cut, a polyphenol oxidase is released which catalyses the oxidation of these phenols to brown polymers. There has been considerable research into ways of reducing the post-harvest browning of apple products, for example, in fruit desserts, particularly since the use of sulfites has been stopped. Application of a dilute solution of a competitive inhibitor of polyphenol oxidase, such as 4-hexylresorcinol, and an antioxidant, such as isoascorbic acid, has been considered. Apples naturally contain variable amounts of vitamin C, particularly in the peal. When apples segments were preserved in Kilner jars, they were blanched for 2–3 minutes to deactivate the polyphenol oxidases. In parts of the West

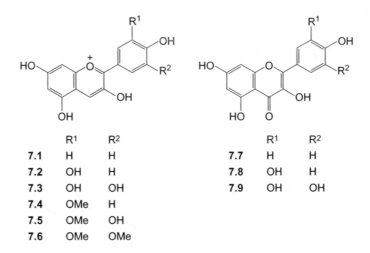

	R¹	R²			R¹	R²
7.1	H	H		7.7	H	H
7.2	OH	H		7.8	OH	H
7.3	OH	OH		7.9	OH	OH
7.4	OMe	H				
7.5	OMe	OH				
7.6	OMe	OMe				

Country, some varieties of apples were pickled for storage by a brief heating in sweetened vinegar and eaten like pickled onions with cheese.

Much of the current work points to the value of the peel. The variation in the pigmentation of the peel is due mainly to the presence of the anthocyanin, cyanidin **7.2** 3-galactoside (red), and the flavonol, quercetin **7.8** 3-galactoside (yellow), against a green chlorophyll background. The minor pigments include the C-3 and C-7 arabinosides of cyanidin **7.2**. There are five major groups of polyhydroxyphenols (hydroxycinnamates, dihydrochalcones, flavanol, flavonols and anthocyanins) with valuable antioxidant activity that are found in apple peel. Together with vitamin C, they make a significant contribution to the biological properties of apples. Phloretin **7.11** and its 2'-O-β-glucopyranoside, phloridzin, are dihydrochalcones which are associated with these beneficial effects. Antioxidant properties are also linked with the presence of oligomeric proanthocyanidins, such as leucocyanidin.

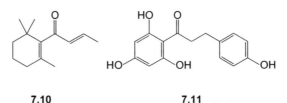

7.10   7.11

Whilst the polyphenols of apple peel may be associated with their antioxidant activity, bioactivity fractionation of extracts from apple peel has revealed that the triterpenoid content is responsible for an antiproliferative tumour inhibitory activity. The common pentacyclic triterpenes, ursolic **7.12** and oleanolic **7.13** acids, and their 2α-hydroxy derivatives, together with their coumaroyl esters have been isolated.

The 2α-hydroxyursolic acid and 2α-hydroxyoleanolic acid (maslinic acid), which were present in 0.01% concentrations, were shown to be active against human liver, colon and breast cancer lines. The epicuticular wax coating of the apple contains long chain hydrocarbons and fatty acids as well as some fatty acid esters of *p*-coumaroyl alcohol.

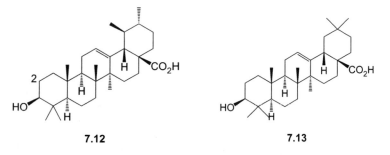

7.12   **7.13**

As with many fruit, the seeds of apples are cyanogenic and contain the cyanogenic glycoside, amygdalin **7.14**. The gibberellin plant hormones that are found in apple seed have been identified by gas chromatography-mass spectrometry at different developmental stages. They include gibberellins with and without a 13-hydroxyl group, whilst others had hydroxyl groups at C-1, C-2, C-3, C-11, C-12 and C-15. Some have an extra ring between C-9 and C-15 (*e.g.***7.15**)

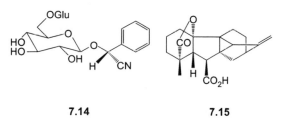

7.14   **7.15**

A number of fungi are found growing on apples, particularly in storage. A pink rot is often *Trichothecium roseum*. This fungus produces two sets of metabolites, the sesquiterpene, trichothecin **7.16** which imparts a bitter taste to the apple, and the diterpenoid rosane lactones, such as rosenonolactone **7.17**. The more highly hydroxylated trichothecenes are serious mycotoxins. Another problem on apples is caused by a bluish-grey mould, *Penicillium expansum* which produces a soft brown rot, often starting from the core of stored apples. Patulin **7.18**, which is mutagenic, is a metabolite of this fungus and it can contaminate apple juice. A maximum residual level of 50 μg kg^{-1} has been recommended. A brown rot with grey or yellow pustules of conidia may be caused by a *Sclerotinia* species, *e.g. S. fructigena*. The infected fruit gradually shrivel as their structure is broken down and they dehydrate.

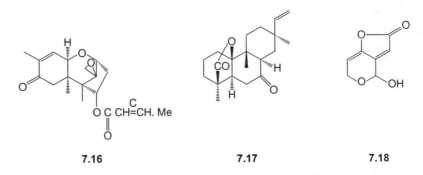

7.16   **7.17**   **7.18**

The larvae of the codling moth, *Cydia pomonella,* can cause serious damage to apples. Seasonal variations in the volatiles from the apple may repel the moth in the early stages of fruit formation up to late May but they then become attractants as the fruit ripens in July and August. When an apple is infected with the larvae of the European sawfly, it emits significant amounts of *trans,trans*-α-farnesene **7.19** and β-ocimene **7.20**. These provide a chemical clue for a parasitic wasp, *Lathrolestes ensator*, to locate the sawfly larvae in which to lay its eggs.

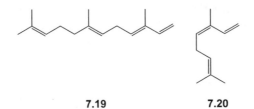

7.19   **7.20**

The flavour of pears (*Pyrus communis*) is associated with the presence of the fruit esters, butyl and hexyl acetate, ethyl butanoate and ethyl (+)-2-methylbutanoate together with the methyl and ethyl esters of (2*E*,4*Z*)-2,4-decadienoic acid.

7.3   QUINCES

The quince (*Cydonia oblonga*) also belongs to the Rosaceae and is related to apples and pears. The fruit are pome fruit and have a structure like that of apples and pears. The fruit is usually too hard to eat raw but is cooked and often used in jam. When quince is cooked, it produces a powerful sweet odour which has been attributed to *cis-* and *trans*-marmelolactone **7.21**. The major volatile compounds obtained from quince include hexanol, *trans*-α-farnesene and farnesol, together with a series of ionone derivatives including theaspirone **7.22** and some isomers of megastigma-4,6,8-triene-3-one **7.23**. A tetrahydrofuran marmelo-oxide related to marmelolactone and quince oxepine **7.24**, has also been isolated. Marmelolactone and the quince oxepine, together with the ionone derivatives, are formed by the degradation of carotenoids. The marmelolactones arise from the central portion.

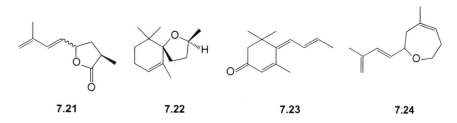

| 7.21 | 7.22 | 7.23 | 7.24 |

Citric, malic and quinic acids, together with antioxidant polyphenols such as chlorogenic acid **7.25** and quercetin **7.8** 3-rutinoside, have also been obtained from quinces. The fruit also contains vitamin C as an antioxidant. Like other fruits of this family, the pips are cyanogenic.

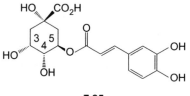

7.25

7.4 PLUMS

Plums are amongst the more widely grown stone fruit trees. The European plum, *Prunus domestica,* belongs to the Rosaceae family and it originated from the Caucasus. The American or Japanese plum (*P. salcinia*) is a closely related species. There are a number of variants such as greengages, whilst some cultivars are used specifically for drying as prunes. Plums are closely related to damsons (*P. institia*), sloes (*P. spinosa*) and apricots (*P. armeniaca*) and there is a hybrid between the apricot and the plum which is known as the 'plumcot'. Plum brandy is known as slivovitz.

The antioxidant components of plums which are of value as inhibitors of the oxidation of low density lipoprotein include vitamin C, (typically 60–250 mg per kg), vitamin E (α- and γ-tocopherols), β-carotene, and glycosides of kaempferol **7.7** and quercetin **7.8** (20–52 mg per kg) such as rutin (quercetin 3-rutinoside). The major antioxidative hydroxycinnamate which is present is neochlorogenic acid (3-O-caffeoylquinic acid, 540 mg per kg). Its isomers, chlorogenic acid **7.25** (5-O-caffeoylquinic acid, 73 mg per kg) and cryptogenic acid (4-O-caffeoylquinic acid, 9 mg per kg) have also been identified. Free caffeic, protocatechuic,

coumaric and ferulic acids have also been detected. Cyanidin **7.2**
3-rutinoside is the major anthocyanin (>60%). It is accompanied by
smaller amounts of the peonidin **7.4** derivative.

Over 70 volatile compounds have been identified in the juices of the
Victoria plum. Benzaldehyde, linalool, ethyl nonanoate, methyl cinnamate
and γ-decalactone contribute to the plum aroma. A number of C_6
aldehydes, alcohols and esters derived from the oxidation of linoleic acid
are also present. These include (*Z*)-hex-3-enal and the corresponding
alcohol, (*E*)-hex-2-enal, hexenyl acetate, hexenyl butanoate and hexenyl
hexanoate. Apart from linalool, other monoterpenes, such as α-terpineol
and borneol, and some hydroxylated linalool derivatives were also detected. A series of hydroxylated and ketonic C_{13} norisoprenoids derived from
the oxidative degradation of carotenoids have also been observed. Some of
these were stored as their glycosides. The release of the aglycones during the
storage and cooking of fruit has a marked impact on their overall flavour.

A characteristic feature of plums is a wax bloom on the fruit. The
epicuticular wax coating on the surface of the plum can act as a reservoir
for compounds, such as nonanal, which contribute to the flavour of the
plum. The wax itself comprises long chain alkanes (20%, mainly C_{29})
and alcohols (48%, mainly C_{29}-10-ol), as well as small amounts of the
triterpenes, ursolic **7.12** and oleanolic **7.13** acids (1.0 and 5.4%,
respectively). The wax coating provides protection against dehydration
and fungal attack.

Typical of stone fruits, plum stones are cyanogenic and contain prunasin (D-mandelonitrile glucoside) and amygdalin **7.14** (D-mandelonitrile
gentiobioside). The bark of the tree contains fraxinol **7.26** (5,7-
dimethoxy-6-hydroxycoumarin), phloroacetophenone, 4-O-methylphlo-
roacetophenone **7.27** and domesticoside (2-O-β-D-giucopyranosyl-4-O-
methylphloroacetophenone), whilst the heartwood contains a group
of dihydroflavanols, isosakuranetin **7.28**, dihydrokaempferide **7.29**,
naringenin **7.30**, 8,4′-dimethoxy-3,5,7-trihydroxyflavanone **7.31**, 3-
methoxy-5,7,4′-trihydroxyflavanone **7.32**, 6,4′dimethoxy-3,5,7-
trihydroxyflavanone **7.33** and prudomestin **7.34**.

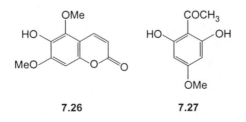

7.26   **7.27**

7.5 PEACHES

The peach (*Prunus persica*) originated in China although, as its name suggests, it was at one time thought to have come from Iran (Persia). The nectarine is a smooth-skinned cultivar. Typical of the Rosaceae, the fruit contains flavanones, such as narigenin **7.30** and persicogenin **7.35**, flavonols such as kaempferol **7.7** and peonidin **7.4** glycosides. Major components of the aroma which have been identified include hexanal,

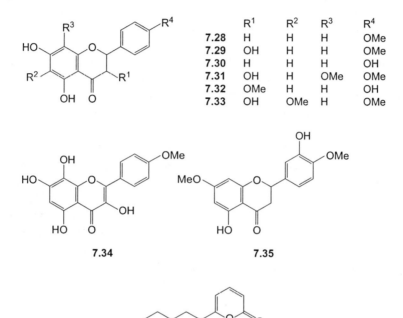

	R¹	R²	R³	R⁴
7.28	H	H	H	OMe
7.29	OH	H	H	OMe
7.30	H	H	H	OH
7.31	OH	H	OMe	OMe
7.32	OMe	H	H	OH
7.33	OH	OMe	H	OMe

7.34   **7.35**

7.36

(*E*)-2-hexenal, benzaldehyde, linalool, 6-n-pentyl-α-pyrone **7.36**, γ- and δ-decalactones, hexadecanoic acid and the marmelolactones **7.21**. The lactones make an important contribution to peach aroma. A number of monoterpenoids, such as geraniol and linalool, and their hydroxylation products and C_{13} norisoprenoids occur as their glycosides. 6-n-Pentyl-α-pyrone **7.36** has antifungal activity. The skins of peaches also produce pentacyclic triterpenes, *e.g.* 1β,2α,3α,24-tetrahydroxyolean-12-en-28-oic acid, as phytoalexins. Typical of the *Amygdali* members of the Rosaceae, peaches are cyanogenic and produce amygdalin **7.14**. The gibberellin plant hormones of the peach have been thoroughly investigated. The sequences of the hormones that were isolated possessed a 13-hydroxyl group and a 1,2-double bond, as well as a hydroxyl group at C-12.

7.6 SWEET CHERRIES

The development of dwarfing root-stock over the last thirty years has meant that it is not only possible to grow cherries (*Prunus avium,* sweet cherry or *Prunus cerasus,* sour cherry) in an average sized garden, but also to protect the small trees against frost damage and to prevent birds from stealing the crop. A number of cultivars are also self-fertile.

The major sugars that are found in sweet cherries are glucose (60–120 g per kg), fructose (48–102 g per kg), sucrose and sorbitol. The main organic acids are malic acid (3.5–8 g per kg) and citric acid (0.1–0.5 g per kg) together with small amounts of shikimic and fumaric acids. The main phenolic acids were neochlorogenic acid (4.7–11.9 mg per 100 g), *p*-coumaroylquinic acid (0.77–7.20 mg per 100 g) and chlorogenic acid **7.26** (0.6–2.6 mg per 100 g). Epicatechin and rutin were detected in the range 0.4–4.5 mg per 100 g and 2.1–5.8 mg per 100 g, respectively. The major anthocyanin pigment was cyanidin **7.2** 3-rutinoside, followed by smaller amounts of cyanidin 3-glucoside, peonidin **7.4** 3-rutinoside and pelargonidin **7.1** 3-rutinoside. Melatonin (*N*-acetyl-5-methoxy-tryptamine) has been detected as an antioxidant in sour cherries.

The volatile compounds emitted by sweet cherries include aldehydes, such as butanal, nonanal, decanal and particularly benzaldehyde, the ketone, 6-methylhept-5-en-2-one, alcohols, such as ethanol, pentan-1-ol and 2-methylpentane-2, 4-diol, and esters, such as ethyl acetate, hexyl acetate, hexyl 2-methylbutyrate and hexyl hexanoate. Sour cherries contain rather more aldehydes. The volatile monoterpenes that are emitted by the cherry vary from flowering to fruiting. Thus α-pinene, camphene and myrcene are major components at the flowering stage, whilst limonene and α-phellandrene become more significant at the fruiting stage. Typical of other stone fruits, cherry stones are cyanogenic and contain amygdalin and prunasin. A range of gibberellins have been identified, some of which contain hydroxyl groups at C-12 and C-15.

The heartwood and the resinous exudate of the cherry tree contain a range of flavonones including techtochrysin, sakuranetin, dihydrowagonin, naringenin, dihydrokaempferol and catechin.

7.7 FIGS

The fig (*Ficus carica*) is a member of the Moraceae family which also includes the mulberry. It has a very long history as a cultivated fruit and it is eaten both fresh and dried. In the Mediterranean it will produce two crops a year, but in temperate climates it only produces a single crop. The fig which we eat is a combination of fruit and flower. The fruit has a high sugar content, as much as 40% of the dried fruit, and it is a good source of minerals such as calcium, iron, magnesium and potassium. A 100 g serving is reported to provide as much as 30% of our daily requirement of iron, 10% of our calcium and 14% of potassium. However, its high fibre and tannin content make it a mild laxative. The fruit can vary in colour and contains the carotenoids, lycopene, lutein and β-carotene, as well as high levels of polyphenol antioxidants. The major anthocyanin that is active in this context is cyanidin **7.2** 3-rhamnoglucoside, which is found in the skin at a concentration of up to 30 mg per 100 g.

Ficus species are pollinated by a specific species of wasp. The failure of figs to develop can be due to the absence of the relevant wasp. The volatile attractants for the fig wasp *Blastophaga psenes*, include linalool, the *cis*- and *trans*-furanoid linalool oxides, the sesquiterpenes, caryophyllene and germacrene-D, and benzyl alcohol. When the fruit is picked or the plant is damaged, it can exude a milky latex. In small doses, this latex has been used as a folk remedy for warts. It contains a cytotoxic 6-O-acyl-β-D-glucosyl derivative of the plant sterol, β-sitosterol. The acyl groups are long chain fatty acids, such as palmitic acid. The sap also contains the furanocoumarins, psoralen **7.37** and bergapten **7.38** which can produce a contact dermatitis.

7.37 R = H
7.38 R = OMe

7.8 BERRY FRUITS

Edible berry or soft fruits are widely grown in the garden. Originally cultivated for their flavour and colourful contributions to fruit dishes, there is now a significant body of research which indicates that edible berries and soft fruit can have a positive effect on human health, including cardiovascular disease, neurodegenerative and other diseases associated with ageing, obesity and certain cancers. Although these fruits contain nutrients such as sugars, typically fructose, glucose and sucrose, vitamins, minerals and dietary fibres, as well as attractive aroma constituents, many of their beneficial biological properties have been associated with their phenolic phytochemical constituents. The phenolic compounds found in berries include phenolic acids and their esters, flavonoids (anthocyanins, flavonols and flavanols), tannins (proanthocyanidins, ellagitannins and gallotannins) and stilbenoids. The concentration of these in the different berry fruits and their variation between cultivars has formed the subject of many studies. Although these phenolic compounds are best known as antioxidants, there is increasing evidence to suggest that they may owe at least some of their biological activity to their regulatory effect on metabolic enzymes, nuclear receptors and cell-signalling pathways. However, it has been pointed out that the concentrations of the berry phytochemicals that have been used in some cell-culture experiments may significantly exceed the levels of these phenolics that are achievable physiologically in humans.

One of the characteristic features of the anthocyanin pigments are the colour changes that occur at different pH, a feature often observed on washing up fruit dishes with washing soda. The changes involve the formation of an anhydro-base and its ionization. Thus cyanidin **7.2**, which is red in acidic solution, is converted to a violet anhydro-base **7.39** which gives a blue salt **7.41** in alkali. At pH 4, the salt cyanidin chloride adds water to form a colourless pseudo-base **7.40** in which the extended conjugated system of the anthocyanin is interrupted. The addition of nucleophiles to the anthocyanins provides an explanation of the use of dilute mildly alkaline hydrogen peroxide in diminishing fruit juice stains.

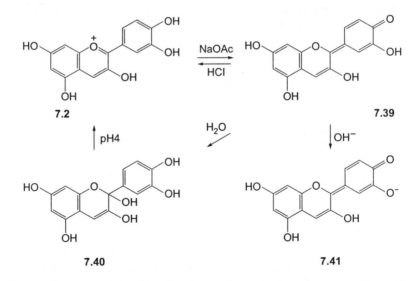

The main organic acids that have been isolated are malic acid and citric acid, whilst a few berries contain benzoic acid, as well as the phenolic acids. The small pips obtained as a by-product from making fruit juices can be a source of unsaturated fatty acids, such as γ-linolenic acid, whilst the presence of some unusual unsaturated nitriles, *e.g.* **7.42**, has been associated with their astringency.

7.9 BLACKCURRANTS, REDCURRANTS AND GOOSEBERRIES

The blackcurrant (*Ribes nigrum*) and its relatives the redcurrants (*R. rubrum*) and gooseberry (*R. uva-crispa*) are popular fruit bushes. They belong to the family Grossulariaceae. The colour of blackcurrants is associated with the presence of cyanidin **7.2** and delphinidin **7.3** glycosides, such as the 3-glucoside, 3-arabinoside, 3-rutinoside, 3-xyloside and their esters. Redcurrants utilize mainly cyanidin **7.2** glycosides as pigments. Blackcurrants contain higher concentrations of oligomeric proanthocyanidins than redcurrants and gooseberries. The hydroxycinnamic acids, caffeic acid and *m*-coumaric acid, as well as *p*-hydroxyphenylacetic acid, protocatechuic acid **7.43** and salicylic acid were identified as both the free acids and bound as esters and as parts of the glycosides in blackcurrants. The flavonoids which have been detected in the leaves and in the seeds include the 3-glucosides and rutinosides of kaempferol **7.7**, quercetin **7.8** and myricetin **7.9**. The vitamin C content of blackcurrants (72–191 mg per 100 g) is quite high.

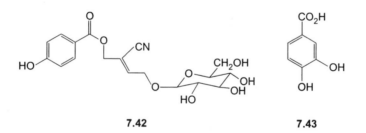

7.42   **7.43**

Blackcurrant aroma is complex. The major components of blackcurrant flavour are esters such as ethyl butanoate, terpenes such as linalool **7.44** 1,8-cineole **7.45**, α-terpineol **7.46**, citronellol and β-damascenone **7.47**. Diacetyl and the C_6 aldehydes such as (*E*)-2-hexenal and the corresponding alcohol are also present, as well as pyrazines, such 2-isopropyl-3-methoxypyrazine. A characteristic feline odour is often observed both on pruning blackcurrant bushes and on harvesting the fruit. This arises from 4-methoxy-2-methylbutane-2-thiol **7.48** which, despite its major olfactory contribution to the aroma, is present at less than 0.1% of the essential oil. The major components (66–75%) of the essential oil of blackcurrant buds are the common monoterpene hydrocarbons including the α- and β-pinenes **7.49**, sabinene **7.50**, $Δ^3$-carene **7.51**, limonene **7.52** and its double bond isomers, β-phellandrene and terpinolene. Smaller amounts of sesquiterpenes such as caryophyllene **7.53**, germacrene B **7.54**, γ-elemene **7.55** and bicyclo-germacrene **7.56** were detected, Terpinen-4-ol, citronellyl acetate, bornyl acetate, spathulenol and caryophyllene oxide were amongst the oxygenated components.

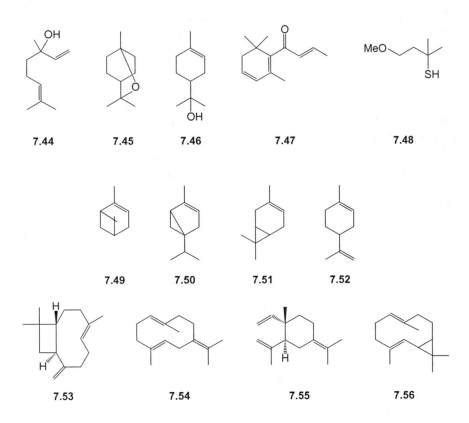

7.44	**7.45**	**7.46**	**7.47**	**7.48**

7.49	**7.50**	**7.51**	**7.52**

7.53	**7.54**	**7.55**	**7.56**

The seeds and buds of blackcurrent are cyanogenic and contain some unusual hydroxynitriles including 3-hydroxy-2-methylbutyronitrile and the (*E*) and (*Z*) isomers of 2-hydroxymethyl-2-butenonitrile. Nigrumin-5-*p*-coumarate and 5-ferulate are 4-glucosides of these 2(*E*)-butyronitriles. The corresponding 4-hydroxy- and 4-hydroxy-3-methoxybenzoates have been isolated from redcurrants. The 3-glucopyranoside of 3,4-dihydroxy-7, 8-dihydro-β-ionone **7.57** and β-ionol have been isolated from the leaves of the redcurrant and the gooseberry. Gooseberry leaves also contain the α-hydroxynitrile glucosides, lotaustralin **7.58** and its epimer, as well as the unsaturated but-2-enenitrile, rhodiocyanoside E **7.59**. These compounds are biosynthesized from isoleucine.

3-Carboxymethylindole-1-*N*- β-D-glucopyranoside **7.60** has been isolated from redcurrants and shown to induce a mouth-drying astringency at very low threshold concentrations of 1 nmol l^{-1}. Other astringent compounds which have been detected in redcurrant juice include rubrumin [(3E,5E)-6-(3-hydroxy-4β-D-glucopyranosyloxy)-phenyl-3,5-hexadiene-2-one] **7.61** and its 3,4-dihydro derivative.

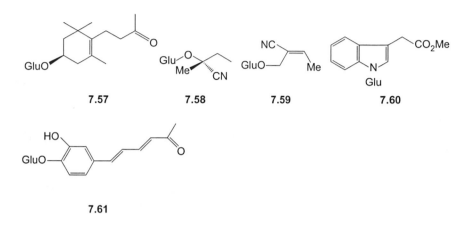

7.57   **7.58**   **7.59**   **7.60**

7.61

7.10   RASPBERRIES

The raspberry (*Rubus idaeus*) belongs to the Rosaceae family. Their fruit may be produced either in the summer or the autumn depending on the cultivar. They contain a variety of beneficial compounds. The principal sugars are fructose (2.35 g per 100 g), glucose (4.86 g per 100 g) and sucrose (0.2 g per 100 g). They also contain vitamin C (26 mg per 100 g) and small amounts of other vitamins. Raspberry polyphenols consist primarily of anthocyanins, principally cyanidin **7.2** 3-O-glycosides (approx. 80 mg per 100 g), and hydrolysable tannins, particularly ellagitannins (300 mg per 100 g). The major glycosides in these include the sophoroside, glucoside, rutinoside and glucosylrutinoside. Red raspberries contain other polyphenols including kaempferol **7.7** and quercetin **7.8** 3-glucuronides and phenolic acids including gallic acid and some complex ellagitannins, such as sanguiin H-6. The ellagic acid conjugates involve glucose esters of the ellagic acid and gallic acid.

Raspberry aroma is a complex mixture. The dominant compounds include 4(4-hydroxyphenyl)butan-2-one (raspberry ketone; rheosmin) **7.62** together with α- and β-ionone, (*Z*)-3-hexenol, β-damascenone **7.47**, linalool, geraniol, sotolone **7.63**, 1-hexen-3-one, 1-octen-3-one and 1-nonen-3-one. The susceptibility of various cultivars to the grey mould, *Botrytis cinerea*, is inversely correlated to the production of various mono- and sesquiterpenes which appear to impede its development.

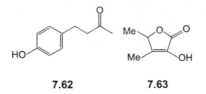

7.62   **7.63**

7.11 STRAWBERRIES

The cultivated garden strawberry, *Fragaria x ananassa*, is a hybrid of two American species, *F. virginiana* and *F. chiloensis*. They belong to the Rosaceae family. This hybrid was actually produced in Europe in the 18th century. Today there are many popular cultivars. During the Middle Ages, the leaves and small fruit of the wild strawberry, *F. vesca*, had both a culinary and medicinal role. One of the herbalists, Culpepper, is reported as describing the plant as being 'singularly good for the healing of many ills'. It is claimed that the strawberry gets its name from a description of the growth of the wild strawberry in which the runners appear to be 'strewn' across the ground. In botanical terms, the fleshy part of the berry which we commonly call the 'fruit' is an accessory fruit and it is not the true fruit derived from the ovaries of the plant. The real fruit are what are commonly called the 'seeds'.

The development of the fruit is under hormonal control and both auxins and gibberellin plant hormones have been implicated in fruit set and development. Twenty different gibberellin plant hormones have been detected. Their structures indicate that an early 12- and 13-hydroxylation (GA_{12}, GA_{53}, GA_{44}, GA_{19}, GA_{20}, GA_1, GA_8 and GA_{29}) (**7.64–7.66**) and a 12,13-hydroxylation (GA_{123} **7.67**, GA_{124} **7.68**, GA_{125} **7.69** and GA_{77} **7.70**) pathway exist.

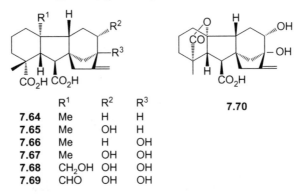

	R^1	R^2	R^3
7.64	Me	H	H
7.65	Me	OH	H
7.66	Me	H	OH
7.67	Me	OH	OH
7.68	CH_2OH	OH	OH
7.69	CHO	OH	OH

7.70

The red colour of the ripe strawberry comes from the anthocyanin, pelargonidin **7.1** 3-glucoside, with cyanidin **7.2** glucosides contributing to the deeper reds. An unusual carboxypelargonidin has also been isolated. The main flavonols are glucosides and glucouronides of quercetin and kaempferol. As the strawberry develops there are two distinct phases of flavonoid biosynthetic activity. The first peak of biosynthetic activity corresponds to the formation of 3',4'-hydroxylated flavan-3-ols, catechin **7.71** and epicatechin **7.72**, and their related proanthocyanidins. Their presence can restrict the growth of the plant pathogenic fungus, *Botrytis cinerea*, on strawberries at this stage in their development. However, these compounds and the proanthocyanidins, such as leucopelargonidin, may contribute to the astringent taste of the unripe fruit. The second peak of biosynthetic activity corresponds to the ripening of the fruit and to the formation of the anthocyanins and flavonols of the red colour.

Strawberries are a useful dietary source of ellagic acid **7.73** and the ellagitannins. The latter are complex molecules which consist of a central core of glucose esterified with hexahydroxydiphenic acid. The hexahydroxydiphenic acid arises by a phenol coupling of gallic acid (3,4,5-trihydroxybenzoic acid).

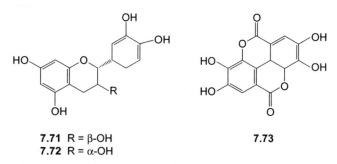

7.71 R = β-OH
7.72 R = α-OH

7.73

Apart from providing the colour of the strawberries, the anthocyanins and flavonols also contribute to the antioxidative potential of the strawberry. The presence of these compounds may offer some protection against cardiovascular disease, inflammation, neurodegenerative diseases and some cancers. There is data which shows that a strawberry extract induced apoptosis in a human colon cancer cell line that expressed the oxidative enzyme, cyclooxygenase-2. This enzyme system is responsible for the biosynthesis of the prostaglandins which are mediators of inflammation. Chronic inflammation predisposes the colon to carcinogenesis.

An interesting aspect of the antioxidant activity of strawberries is the ability of some constituents to inhibit cytochrome P_{450} monooxygenases. A number of these enzymes, such as CYP3A4, play a major role in the oxidative metabolism of compounds in the liver. It has been suggested that as many as 50% of medicines are metabolized by CYP3A4. Several foods and herbs which inhibit CYP3A4 activity, affect drug metabolism. The *cis* and *trans* isomers of kaempferol **7.7** 3- β-D-(6-*p*-coumaroyl)-glucopyranoside obtained from strawberries, significantly inhibited the human CYP3A4 enzyme in an assay involving the oxidation of the heart drug, nifedipine. This suggests that consumption of strawberries might reduce the 'first-pass loss' of some drugs and increase the amount reaching the target organ.

Over 300 volatile compounds have been detected in strawberry flavour, of which 15–20 are considered to be essential for the sensory quality of the strawberry. Key components of the mixture are 2,5-dimethyl-4-hydroxy-3(2H)-furanone and its methyl ether (mesifuran) **7.74**, which have been shown to be derived from glucose *via* fructose. The parent hydroxy compound (Furaneol®) has a caramel-like odour when it is concentrated. Various fruit esters, such as methyl and ethyl butanoate, ethyl 2- and 3-methylbutanoate, methyl hexanoate and the *cis* and *trans* isomers of 2-hexenyl acetate, together with the hexenols are present. The aldehyde, hex-2-enal, which is formed when the strawberry is damaged, is the common leaf aldehyde associated with the smell of mown grass. Terpenoids can reach up to 20% of the volatile material. Monoterpenes, such as linalool, and sesquiterpenes, such as nerolidol, are present.

7.74

These terpenes can affect the attack of a number of fungal and insect pests. The strawberry responds to fungal attack by producing phytoalexins. The pentacyclic triterpenes, euscaphic acid **7.75**, tormentic acid **7.76** and pyrianthic acid **7.77**, have been identified in this context and have been shown to be active against *Botrytis cinerea* and *Colletotrichum fragariae*.

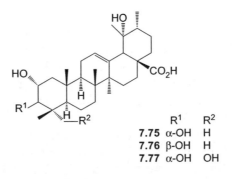

	R¹	R²
7.75	α-OH	H
7.76	β-OH	H
7.77	α-OH	OH

7.12 GRAPES

Grapes (*Vitis vinifera*) have been grown as a fruit for eating and for the production of wine since Biblical times. There are many varieties of grape particularly amongst those that are used for making wine. These include white (green) and red (purple) varieties. The constituents of grapes and the wines that are produced from them have been very thoroughly investigated. It has been estimated that over 800 compounds have been detected in the grapes and various wines. Indeed, it is not always clear whether a compound is a constituent of the parent grape or has been formed as a consequence of fermentation. As the grape develops, there is a transition from berry growth to berry ripening which is accompanied by a change in colour. This change is known as 'veraison'. In the period between veraison and maturity, the acidity of the grape decreases as malic acid is degraded, leaving tartaric acid as the major fruit acid together with citric acid. Sugars, particularly glucose and fructose, accumulate and the concentration of sugars can become quite high. Proline and arginine are the predominant amino acids that are present, together with smaller amounts of glutamic acid, glutamine and alanine.

The flavour of the grape includes a subtle blend of monoterpenes, C_{13} norisoprenoids and fruit esters, such as ethyl 3-methylbutanoate. Although carotenoid levels are quite low in the mature grape (*ca.* 0.8–2.5 mg per kg) their degradation to C_{13} norisoprenoids makes an important contribution to the organoleptic properties of grapes. The major carotenoids that are present are β-carotene, lutein and neoxanthin, with smaller amounts of flavoxanthin, lutein 5,6-epoxide, luteoxanthin and violaxanthin. Degradation of these can give rise to different C_{13} norisoprenoids, which include damascenone **7.47**, 3-hydroxy-β-damascanone **7.78**, 3-oxo-α-ionol, β-ionone, vitispirane **7.79**, and 1,1,6-trimethyl-1,2-dihydronaphthalene. 2,6,6-Trimethylcyclohex-2-ene-1,4-dione has been detected as a minor component. A number of these compounds are found in the grape as their glycosides. The chemistry of the decomposition of the metabolite, 3,4,9-trihydroxymegastigma-5,7-diene, to (*E*)-1-(2,3,6-trimethylphenyl)buta-1,3-diene has been studied. A number of monoterpenes such as linalool, *cis* and *trans* furanolinalool oxide, geraniol, nerol and some of their 6,7, and 8-hydroxylated derivatives, have been detected in the free state and as their glycosides. A decrease in the amount of the highly odorous compound, 2-methoxy-3-(2-methylpropyl)pyrazine, during maturation has been described. The pyrazine ring is probably formed from amino acids.

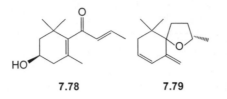

7.78   **7.79**

The anthocyanin colouring matter of the skin depends on the variety but includes the 3-glucosides of the methyl ethers, peonidin **7.42** and malvidin **7.6**, together with cyanidin **7.2** and delphinidin **7.3**.

Two constituents of the skin are the hydroxystilbenes, resveratrol **7.80** and its oligomeric derivative, viniferin **7.81**. Resveratrol is a phytoalexin and possesses antimicrobial properties. It is found in red wine and is a powerful antioxidant which confers considerable health benefits. The French have a similar consumption of dietary fat but a much lower incidence of coronary heart disease when compared with people in the UK. This difference was traced to the higher consumption of red wine by the French and eventually to the presence of resveratrol arising from the red grape skins. The cardioprotective effect of resveratrol was associated with its antioxidant activity. By scavenging reactive oxygen species, resveratrol inhibits the peroxidation of low density lipoprotein and platelet aggregation. It also inhibits the enzyme 5-lipoxygenase which

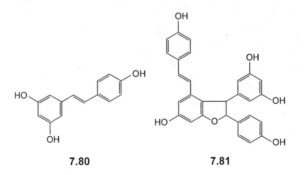

7.80   **7.81**

mediates a key step in the biosynthesis of the leukotrienes from arachidonic acid. This confers some anti-inflammatory activity. Resveratrol also possesses some tumour inhibitory activity.

CHAPTER 8
Culinary Herbs

8.1 INTRODUCTION

Culinary herbs, whether they are grown on the kitchen window sill, on the patio or in a formal part of the garden, make a useful contribution to edible chemistry. Before the advent of many synthetic medicines, herbal medicines played an important role in the health of the nation. In the past, the medicinal value of the herbs often far outweighed their culinary usage. Indeed, a number of their bioactive components have provided the starting point for the development of synthetic medicines. Prior to the invention of refrigerators, the antioxidant and antimicrobial biological activity of the constituents of herbs were useful in slowing the bacterial spoilage of food, whilst their aromatic properties masked the odour and taste arising from any spoilage. Most of the natural products that contribute to the aromatic properties of culinary herbs are mono- and sesquiterpenes and simple phenylpropanoid metabolites. These may also be stored in the plant as glycosides and the free aglycone released by enzymatic hydrolysis when the cells are disrupted.

A number of these natural products have insecticidal and allelopathic activity – functions which can be of benefit but also produce limitations on where the plants can be grown. Some like chives, a member of the onion family, are small plants which can be grown amongst other established vegetables so that the insecticidal and allelopathic activity may protect the main crop. Many of the terpenes are chiral molecules and their particular odour is associated with a specific enantiomer. Thus the R-(−)-enantiomer of carvone **8.1** which occurs in mint has a

Chemistry in the Kitchen Garden
By James R. Hanson
© James R. Hanson 2011
Published by the Royal Society of Chemistry, www.rsc.org

peppermint smell, whilst its S-(+)-enantiomer **8.2** which occurs in caraway, dill and fennel has a different smell.

The antioxidant and antibacterial activity of herbs can often be attributed to their phenolic content. These compounds include phenolic acids such as rosmarinic acid, as well as flavonoids and some diterpenoid products. These compounds, together with some higher molecular weight tannin oxidation products, can also have an anti-inflammatory action.

Many of the common plants that are used as culinary herbs belong to the Lamiaceae and Apiaceae (Umbelliferae) families and will be grouped together in this order.

8.2 MINTS

Mint is one of the commonest culinary garden herbs. A sprig of mint is often added to many dishes including potatoes and peas and even to soups and to some sweets and ice-cream, whilst mint sauce is a regular accompaniment to lamb. Like many useful herbs, *Mentha* species belong to the Lamiaceae (Labiatae) family. However, taxonomically the situation is quite complex and there are many variants and different cultivars of species within this genus. The best known garden species are spearmint (*M. spicata*) and peppermint (*Mentha x piperita*). The latter is a hybrid of *M. spicata* and *M. aquatica*. Applemint, *Mentha rotundifolia,* is sometimes considered as a hybrid of *M. longifolia* and *M. suaveolens.* Pennyroyal is *M. pulegium,* whilst cornmint is *M. arvensis.* Catmint, on the other hand, is a different species, *Nepeta cataria.* These different species coupled with the existence of different chemotypes and the tendency of these plants not to come true to seed, make it difficult for the gardener to select a mint to grow. The best way forward is to select a plant from a nursery or a friend based on smelling the leaves first. Fortunately, mint is easy to propagate by striking cuttings. Although it will grow in a variety of soils, it is best in a deep loose sandy soil and typical of a Mediterranean plant, in a good sunny position. However, it is a very invasive plant, spreading with creeping roots dominating its competitors with the aid of allelopathic monoterpenes. In a small garden, it is best constrained by growing it in a good sized pot. The common *Mentha* species are perennials which in the UK die back outdoors over the winter months but the cut leaves preserve well.

The variety of mints produces a wide range of organoleptic components which affect their culinary usage. The majority of the monoterpenes that are produced by the mints are oxygenated *p*-menthane derivatives. The essential oil from *M. spicata* is often rich in

R-(−)-carvone **8.1** and also contains variable amounts of limonene **8.3**, piperitone **8.4** and its epoxide, menthone **8.5**, isomenthone and pulegone **8.6**. On the other hand, *M. x piperita* contains limonene **8.3** and menthone **8.5** as the major monoterpenes, together with a small amount of pulegone **8.6** and traces of the sesquiterpene hydrocarbons, humulene **8.10**, caryophyllene **8.11** and germacrene D **8.12**. Pulegone **8.6** is the major constituent of *M. pulegium* where it is accompanied by a smaller amount of isomenthone. Piperitone epoxide is a major component of the essential oil of *M. x rotundifolia*. Menthofuran **8.7**, mintlactone **8.8** and isomintlactone have also been obtained from this species. *M. arvensis* is sometimes grown as a source of menthol **8.9**.

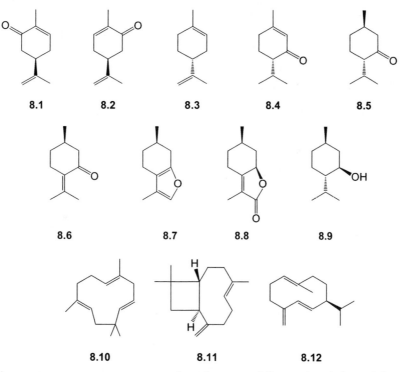

These monoterpenes are stored and secreted from glandular trichomes in the leaves. The alcohols may also be stored as their glycosides and released by enzymatic hydrolysis when the leaf is disrupted. Crushing the leaves between the fingers not only has the effect of disrupting the oil glands but also of bringing a glycosidase into contact with its substrate.

The mint monoterpenes, particularly pulegone **8.6** from pennyroyal oil, are effective insecticides. However, there is evidence that they may act synergistically rather than as individual compounds. It has been suggested that mint will deter the growth of aphids on adjacent plants. The monoterpenes from mint have a significant allelopathic effect preventing the germination and spread of other plants. If the soil from a bed of mint is crushed, it will easily reveal the smell of the monoterpenes. Once the mint is growing during the summer, competitive weeds are less common.

Monoterpenes are not the only biologically active constituents of mints. *M. spicata* contains a range of phenolic antioxidants which can have a preservative action in foodstuffs. The use of mint sauce with lamb may have its origins in reducing the rancidity of the animal fats. The antioxidants include caffeic acid **8.13**, rosmarinic acid **8.15** and the flavonoids, eriocitrin **8.16**, luteolin **8.18** 7-O-glucoside and isorhoifolin **8.19**. These compounds not only possess radical scavenging ability but also complex with iron, which might otherwise participate in radical-generating Fenton type oxidative reactions. Certain flavones, particularly thymonin **8.20**, have shown antihistamine activity and may account for some of the anti-inflammatory activity of mint preparations. Steroidal glycosides and pentacyclic triterpenes of the ursane series, together with β-carotene and lutein, have been detected.

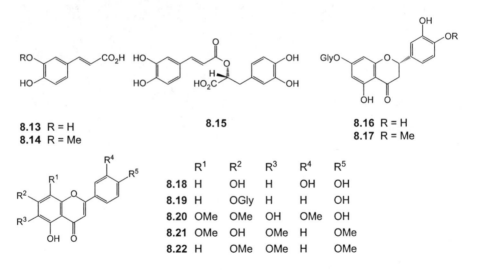

8.13 R = H
8.14 R = Me

8.15

8.16 R = H
8.17 R = Me

	R^1	R^2	R^3	R^4	R^5
8.18	H	OH	H	OH	OH
8.19	H	OGly	H	H	OH
8.20	OMe	OMe	OH	OMe	OH
8.21	OMe	OH	OMe	H	OMe
8.22	H	OMe	OMe	H	OMe

The common catmint or catnip (*Nepeta cataria*) is attractive to members of the cat family. The active component, nepetalactone **8.23**, is an iridoid monoterpenoid which is biosynthesized from geranyl diphosphate by a cyclization that is different from that leading to the *p*-menthane skeleton. When the structure of nepetalactone was established in 1941, the natural product was bioassayed against African lions in Villas Zoo in Madison, Wisconsin. The lions were reported to be aroused from a 'state of lethargy to one of intense excitement' and to become 'ludicrously playful' when given cotton rags soaked in a dilute solution of the lactone. When a sample of the compound was synthesized in the 1960s in a laboratory in Australia, a sample was left in a cupboard in the laboratory over the weekend. There were reports that the laboratory was invaded and damaged by feral cats from the neighbourhood.

8.23

8.3 BASIL

The genus *Ocimum* (Lamiaceae) grows mainly in tropical or sub-tropical regions but basil (*O. basilicum*) can be grown as a greenhouse or indoor culinary herb. The leaves have an aniseed-like aroma and a slightly sweet taste. The main components of the essential oil are compounds derived from the phenylpropanoid pathway and include methyl chavicol (estragole) **8.24** and eugenol **8.25**, as well as the acyclic monoterpenes, citronellol, geraniol and linalool. The antioxidant phenolic acid, rosmarinic acid **8.15**, as well as β-sitosterol and the triterpenes, oleanolic and ursolic acids have also been detected.

A feature of many of the plants of the Lamiaceae including basil is the secretion of a coating on the leaves which contains various flavonoids. The light absorption of these lipophilic methoxylated flavonoids, such as nevadensin **8.21** and salvigenin **8.22**, may give the plant protection against harmful ultra-violet radiation. They also have antibacterial and antifungal activity. Sweet basil also has insect-controlling activity and this has been associated with the presence of an insect juvenile hormone mimic, juvocimene **8.32**. This compound has an unusual C_6-C_3+ monoterpenoid structure.

	R^1	R^2	R^3	R^4
8.24	H	OMe	H	H
8.25	OMe	OH	H	H
8.26	O-CH₂-O		H	H
8.27	OMe	O-CH₂-O		OMe
8.28	OMe	OMe	OMe	OMe
8.29	OMe	OMe	OMe	H
8.30	O-CH₂-O		OMe	H
8.31	O-CH₂-O		OMe	OMe

8.4 THYME

The aromatic properties of thyme (*Thymus vulgaris*) has figured in literature as in the Shakespearean phrase 'bank whereon the wild thyme grows'. The essential oil from thyme has a high phenol content and includes the monoterpenoid phenols, thymol **8.33** and carvacrol **8.34**. These give it a powerful antiseptic action. Thymol **8.33** has been used in dentistry, in mouthwashes and in throat pastilles (glycerine–thymol compound BPC). It has also been used as an expectorant and in the treatment of bronchitis. The isomer carvacrol **8.34** is used in several 'vaporisers' to alleviate the congestion of a cold. A biphenyl, 3,4,3′,4′-tetrahydroxy-5,5′-diisopropyl-2,2′dimethylbiphenyl arising from a phenol coupling of *p*-cymene-2,3-diol has been isolated together with some methylated flavonoids. The common flavonoid, luteolin **8.18**, has also been detected as an antioxidant. Another group of compounds which have been obtained are glycosides of 4-hydroxyacetophenone.

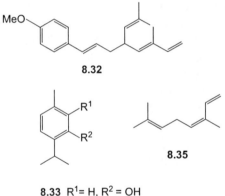

8.32

8.35

8.33 R¹= H, R² = OH
8.34 R¹= OH, R² = H

8.5 LEMON BALM

Lemon balm (*Melissa officinalis*) has been used as a herbal tea to alleviate digestive and sleep disorders. The major components of the essential oil are the aldehydes, citral and citronellal, accompanied by the corresponding alcohols, the hydrocarbon β-ocimene **8.35** and the sesquiterpene hydrocarbons, caryophyllene **8.11** and germacrene-D **8.12**. The leaves also contain caffeic acid **8.13**, rosmarinic acid **8.15** and flavonoids, such as luteolin **8.18** 7-O-glucoside.

8.6 ROSEMARY

The shrub, rosemary (*Rosmarinus officinalis*) is a decorative and aromatic plant which is widely used as a culinary herb. In the past, it had a number of medicinal uses and there are underlying chemical reasons for the saying that 'rosemary is for remembrance'. There is some evidence that certain constituents, particularly the monoterpene, 1.8-cineole **8.36**, and the triterpene, ursolic acid **8.40**, inhibit acetylcholinesterase. A deficiency of acetylcholine in the brain is associated with Alzheimer's disease. The inhibition of acetylcholinesterase can prolong the activity of the limited amount of acetylcholine that is present.

Other monoterpenes that are present include α-pinene **8.37** and camphor **8.38**. The plant also produces a family of diterpenoid phenols and their oxidation products such as carnosol **8.39**, carnosic acid **8.40**, rosmanol **8.41** and some related quinone methides, which have antioxidant, antibacterial and anti-inflammatory activity. Together with common antioxidants such as rosmarinic acid **8.15** and flavonoid glucosides based on hesperidin **8.17** and luteolin **8.18**, these contribute to the preservative role of rosemary in foodstuffs. The diterpenoid carnosic acid, which is also found in sage, has a marked gastroprotective effect in the prevention of gastric ulcers. It inhibits human 5-lipoxygenase and suppresses some pro-inflammatory responses. Rosemary oil not only has a powerful allelopathic effect and inhibits the germination of seeds but it also possesses insecticidal activity. Its constituents have been investigated for use against the spider mite, *Tetranychus urticae*, which is a serious pest in greenhouses.

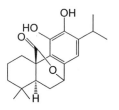

8.36   **8.37**   **8.38**   **8.39**

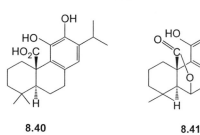

8.40   **8.41**

8.7 HYSSOP

Hyssop (*Hyssopus officinalis*) is a typical member of the Lamiaceae which produces narrow leaves and blue flowers on a spike. It will grow as a perennial shrub in a sunny position but it needs protection over the winter. The flavour is quite strong. The monoterpenes found in the essential oil are β-phellendrene **8.42**, β-pinene **8.43** and the ketones, pinocarvone **8.44**, pinocamphone **8.45**, and isopinocamphone. Some sesquiterpenes including caryophyllene **8.11**, germacrene D **8.12** and hedycaryol **8.46** were also detected. The antioxidant diterpenoid, rosmanol 7-ethylether, has been found in the leaves.

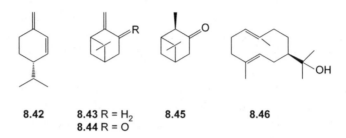

8.42	8.43 R = H₂	8.45	8.46
	8.44 R = O		

Hyssop has a long reputation as a medicinal herb. However, there is evidence that the hyssop which is referred to in the Bible was a carvacrol-producing chemotype of *Origanum* (*Majorana*) *syriaca,* a form of oregano.

8.8 MARJORAM

There are several varieties of marjoram, some of which are also known as 'oregano'. The common type is a sweet marjoram (*Origanum marjorana*) which is related to *O. vulgare* (oregano). Typical of the Lamiaceae, the major aroma constituents are monoterpenes. Sweet marjoram oil contains terpinen-4-ol **8.47**, *cis*-sabinene hydrate **8.48**, *p*-cymene, the α- and γ-terpinenes and carvacrol, as well as geraniol and linalool and their acetates. Oreganum oil contains much higher amounts of carvacrol **8.33** and thymol **8.34**. The presence of these phenols confers antiseptic and preservative properties on this oil. Oregano has antioxidant and anti-inflammatory activity and this has been associated with the presence of the triterpenes, oleanolic **8.49** and ursolic **8.50** acids, and rosmarinic acid **8.15**. The latter is widespread amongst the herbs of the Lamiaceae.

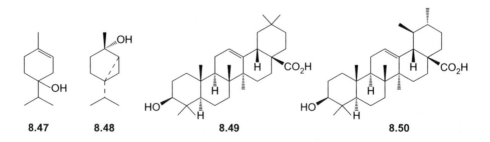

8.47	8.48	8.49	8.50

8.9 SAGE

The sages are another widely grown group of herbs that belong to the Lamiaceae. Sage and onion stuffing is traditionally used with turkey at Christmas. In the days before refrigerators, the antibacterial components of this mixture probably played a role in preserving meat over several days. The name *Salvia*, which is given to plants of this genus, reflects their healing properties. There is an old saying which refers to sage (*S. officinalis*) and reveals the seasonal production of natural products,

> '*He that would live for aye*
> *Must eat sage in May*'

In years gone by, sage was commonly used as a household remedy to control digestive disorders, inflammation, sore throats and nervous headaches.

There are a number of different *Salvia* species that are grown in the garden. These include the bright red, *S. splendens*, and the delicate blue, *S. farinacea*. The leaves of a rare Mexican species, S. *divinorum*, contain a hallucinogenic diterpenoid, salvinorin A **8.51**. This clerodane has attracted interest because it targets the κ-opioid receptor in the brain.

The typical constituents of the essential oil of *S. officinalis* include the monoterpenes, α- and β-pinenes **8.37** and **8.43**, 1,8-cineoie **8.36**, α- and β-thujone **8.52** and camphor **8.38**. The sequiterpenes, humulene **8.10**, caryophyllene **8.11** and viridiflorol, and the diterpene, manool **8.53**, have also been detected. However, there were marked seasonal variations and differences between parts of the plant and between cultivars such as Dalmation sage. It is worth noting that there is some toxicity associated with the presence of thujone and there are regulations which limit the amount that is present in commercial oils.

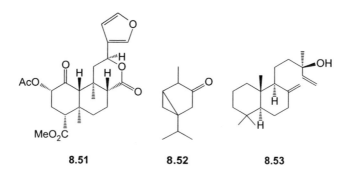

| 8.51 | 8.52 | 8.53 |

Like rosemary, sage has been shown to possess acetylcholinesterase inhibiting properties and to improve mood and reduce anxiety. A series of phenolic diterpenes with an abietane skeleton have been isolated from sage including carnosol **8.39**, carnosic acid **8.40** and rosmanol **8.41**. Carnosol has antioxidant properties and an oxidation product, sagequinone methide A **8.54**, has been isolated. The 6,7-secoabietane, safficinolide **8.55**, and the norditerpene, sageone **8.56**, both showed antiviral activity. The leaves also have anti-inflammatory properties and this has been associated with the presence of the triterpene, ursolic acid **8.50**. The triterpenes, α- and β-amyrin and oleanolic acid, have also been isolated from the leaves.

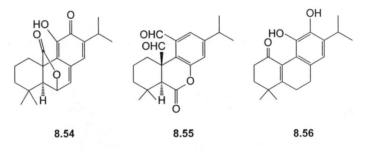

8.54   8.55   8.56

Sage also contains a range of glycosidic phenolic antioxidants derived from the C_6-C_3 pathway. These include rosmarinic acid **8.15**, luteolin **8.18** 7-O-glucoside, the 4-O-glucoside of 4-hydroxyacetophenone (picein) and 5-methoxysalvigenin.

Clary sage is *S. sclarea*. Myrcene, limonene, linalool, geraniol and nerol and their acetates, together with α-terpineol, have been found in collections of this sage. It also contains some sesquiterpenes such as caryophyllene **8.11**, caryophyllene oxide, germacrene-D **8.12** and α- and β-eudesmol **8.58** The diterpene, sclareol **8.57**, is an important constituent and the plant is grown commercially to produce this compound from which various perfumery products are obtained.

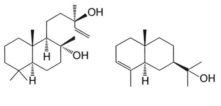

8.57   8.58

This wide range of mostly beneficial compounds suggests a chemical rationale for ranking sage quite highly amongst the culinary herbs.

8.10   PARSLEY

Not all aromatic culinary herbs belong to the Lamiaceae, parsley, dill and fennel belong to the Apiaceae (Umbelliferae). Apart from the ubiquitous terpenes, characteristic metabolites of these plants include furanocoumarins, phenylpropanoids and polyacetylenes. These are compounds which have already been discussed as constituents of carrots, parsnips and celery.

Parsley (*Petroselinum crispum*) is one of the most commonly grown members of this family. There are two major types, a curly-leafed and a flat-leafed variety. A rarer form is a turnip-rooted variety which is grown for its enlarged edible root. The leaves of parsley are used as a culinary decoration and in parsley sauce to go with fish. The plant is grown as an annual. The major monoterpene which is responsible for the organoleptic properties, is *p*-mentha-1,3,8-triene **8.59**. Other monoterpenes that have been found include limonene **8.3**, phellandrene **8.42**, myrcene and the α- and β-pinenes, **8.37** and **8.43**. The seed oil contains rather less of the monoterpenoid constituents and more phenylpropanoids, such as apiole **8.27** and 1-allyl-2,3,4,5-tetramethoxybenzene **8.28**. It also contains a small amount of elemicin **8.29** and myristicin **8.30** which have a nutmeg odour. The fatty acid, petroselinic acid (*Z*-6-octadecenoic acid), was first isolated as a glycerol ester from parsley. The roots contain the C_{17} polyacetylenic alcohols, falcarinol and falcarindiol which are found in other members of the Apiaceae. Furanocoumarins, such as psoralen **8.60**, bergapten **8.61**, xanthotoxin **8.62** and oxypeucadin **8.63** which are also typical products of this family, have been found in parsley. Exposure to these photoactive compounds can produce a contact photodermatitis.

8.59

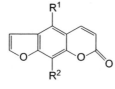

8.60 R¹= R²= H
8.61 R¹= OMe, R²= H
8.62 R¹= H, R²= OMe

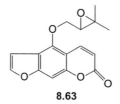

8.63

8.11   CORIANDER

The fruit of coriander (*Coriandrum sativum*) provide an aromatic spicy taste and when dried, form part of curry powders. Fresh coriander leaves are also occasionally used as a garnish and can be known as 'cilantro' or 'Chinese parsley'. However, they have quite a strong odour. The essential oil contains linalool, nerol, *p*-cymene, α-pinene and camphor, whilst a number of furoisocoumarins, such as coriandrin **8.64** and the coriandrones *e.g.* **8.65**, have been isolated from the plant. The latter retains the complete isoprene unit. The fruit also contain the furocoumarin, bergapten **8.61** and its 8-methoxy isomer, xanthotoxin **8.62**. These furanocoumarins and furanoisocoumarins are photoactive and along with linalool, they can produce an allergic reaction. Coriander oil also has a significant antibacterial activity.

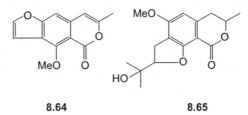

                    8.64   **8.65**

8.12   DILL AND FENNEL

Dill oil (*Anethum graveolens*) contains (+)-limonene **8.66**, (+)-carvone **8.2**, the *cis* and *trans*-carveols **8.67** and dill ether **8.68**, as well as the phenylpropanoid, dill apiole **8.31**. The (+)-carvone has been used as a commercial sprouting inhibitor for potatoes.

Fennel (*Foeniculum vulgare*) is quite a tall perennial member of the Apiaceae. Its chopped leaves are used as a culinary herb. It should not be confused with a relatively uncommon vegetable which is sometimes given the same name. The main components of the essential oil are the phenylpropanoids, estragole **8.24**, and its double bond isomer, anethole, together with the monoterpenoids, limonene and fenchone **8.69**. *p*-Anisaldehyde and α-phellandrene have also been detected. The oil contains the unsaturated fatty acid, petroselinic acid (*Z*-6-octadecaenoic acid) which is an isomer of oleic acid. The oil from fennel has been shown to have a useful acaricidal effect against two common household dust mites, (*Dermatophagoides sp.*). The major biologically active constituents in this context were (+)-fenchone and *p*-anisaldehyde.

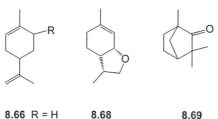

8.66 R = H   **8.68**   **8.69**
8.67 R = OH

8.13 CUMIN

There are two distinct plants with the name cumin that are used to provide a flavouring. Distillation of the seeds of *Cuminum cyminum*, which is a member of the Apiaceae, provides an oil which contains the aldehydes, cuminaldehyde (4-isopropylbenzaldehyde), 1,3- and 1,4-menthadien-7-al **8.70** and 3-menthen-7-al, as well as the monoterpene hydrocarbons, *p*-cymene, myrcene, α- and β-phellandrene, α-terpinene and β-pinene. On the other hand, black cumin, *Nigella sativa,* is a member of the Ranunculaceae (the buttercup family). The oil from its seeds contains *p*-cymene, α-thujene **8.72**, α- and β-pinenes, γ-terpinene, terpinen-4-ol, carvacrol, the *cis* and *trans*-thujan-4-ols, **8.73** and *trans*-4-methoxythujane **8.74**. Whilst the seeds are currently used for flavouring purposes, for example in seed bread, in the past the plants had medicinal uses particularly in the treatment of stomach disorders. At one time, the semicarbazide of cuminaldehyde attracted interest as a potential anti-viral agent. The seeds of caraway, *Carum carvi* (Apiaceae) which contain (+)-carvone and limonene are also used in this context.

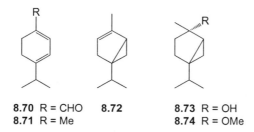

8.70 R = CHO   **8.72**   **8.73** R = OH
8.71 R = Me   **8.74** R = OMe

8.14 BAY LAUREL

Bay laurel (*Laurus nobilis*) is an evergreen tree whose leaves are widely used as a condiment. It belongs to the Lauraceae family and is quite distinct from the garden laurel (*Prunus laurocerasus*). Indeed it is hazardous to use the leaves of the latter as they are cyanogenic. Although sweet bay or bay laurel is a Mediterranean plant, it can be grown in this country in a sheltered position as a patio plant where it may be protected in the winter. The leaves have a wax coating of long chain hydrocarbons. The main components of the essential are 1,8-cineole **8.36**, linalool, α-terpinene **8.71**, α-terpinyl acetate and the methyl ether of eugenol. The flowers contain more sesquiterpenes including caryophyllene **8.11**, β-elemene **8.75** and germacradienol. The roots and leaves are also a source of sesquiterpenoid germacranolide lactones, such as costunolide **8.76** A series of megastigmane glucosides known as the laurosides **8.77** have also been isolated. Like many essential oils, bay laurel oil possesses a strong antibacterial effect, particularly against food borne pathogenic bacteria.

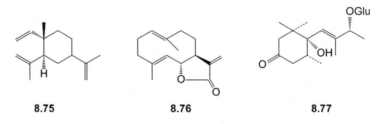

| 8.75 | 8.76 | 8.77 |

The wax coating of the evergreen laurel leaves can absorb atmospheric pollutants. The level of polycyclic aromatic hydrocarbons absorbed into the wax coating of bay laurel has been used as a measure of urban air pollution. This may serve as a warning to those who grow herbs on the patio.

8.15   CHIVES

Chives (*Allium schoenoprasum*) are members of the Alliaceae. Their thin, tubular stems are used as a garnish. Like other members of the onion family (see Chapter 3) their chemistry is dominated by that of their sulfur-containing metabolites. The major flavouring components are methiin (S-methylcysteine-S-oxide), isoalliin (S-(*E*)-1-propenylcysteine-S-oxide) and its dihydro derivative, propiin. The enzyme system alliinase releases the volatile sulfur-containing components which are then transformed into the metabolites (discussed in Chapter 3). When preparing chives, it is important to cut the stems carefully with scissors in order to minimize the contact between alliinase and its substrates before the herbs are used. Although chives are often grown in pots on the patio or windowsill, the presence of these sulfur-containing metabolites and their insecticidal properties, makes them useful in the kitchen garden as a companion plant to protect other vegetables.

8.16   THE CURRY PLANT

The so-called curry plant (*Helichrysum italicum* or *H. angustifolium*) is a widespread Mediterranean medicinal herb which is a member of the Asteraceae. It is sometimes grown in herb gardens, particularly in chalky areas. It has little to do with curries but gets its name because it is highly aromatic. In appearance it has silver leaves which look like lavender. The plant has flavouring, fungicidal and anti-inflammatory properties. The major components of the oil include neryl acetate, nerol, linalool, limonene and α-pinene, together with some sesquiterpenes, such as γ-curcumene, aromadendrene and the eudesmane, rosifoliol. The hydroxylated derivatives of linalool that are present are also found as glycosides. The anti-inflammatory activity has been associated with the presence of acetophenone derivatives and the acylbenzofuran, 12-hydroxytremetone

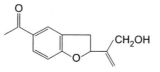

8.78

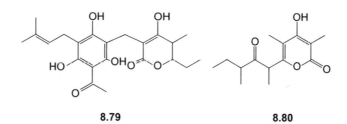

8.79   **8.80**

8.78, together with α-pyrones, such as arzanol **8.79** and micropyrone. **8.80**. Tarragon (*Artemisia dracunculus*) is another member of the *Asteraceae* which is used as a herb. Like fennel, the phenylpropanoid, estragole **8.24** and its double bond isomer, anethole, are major components of the essential oil. Estragole is a suspect carcinogen. The oil also contains a number of monoterpenes including the ocimenes.

In this chapter, we have seen the range of compounds that are found in culinary herbs. Although many are volatile members of the phenylpropanoid and terpenoid families, the diversity of their oxygenation patterns and the complexity of the mixtures that comprise the essential oils of these plants lead to very different organoleptic properties. The origin of many herbs as medicinal plants is reflected in the biological activity of their constituents, indicating that they may possess a beneficial role beyond that of just flavouring foods.

Epilogue

The aim of this book has been to describe the natural products that are found in fruit and vegetables and which contribute to their distinctive characteristics. Over the last fifty years, the immense advances in separation methods and spectroscopic techniques for structure elucidation have led to the identification of a wide range of natural products in fruit and vegetables. Not only have many of their beneficial properties been recognized but also their ecological roles in the development of plants have been identified. The functional role of many of these natural products is to mediate the balance between an organism and its environment in terms of microbial, herbivore or plant–plant interactions.

Nevertheless, many challenging chemical problems await a solution. Analysis of the genome of organisms is starting to reveal that many have a wider biosynthetic potential than that which they currently express. In microorganisms, whole genome sequencing programmes have revealed that the genes encoding biosynthetic enzymes outnumber the known metabolites and secondly that the genes that code for secondary metabolite biosynthetic pathways are often clustered in the microbial genome. The presence of these silent genes has led to the concept of 'cryptic' natural products whose formation may only be elicited by some changes experienced by the organism. In the case of plants, for example, they produce both constitutive antifungal agents and inducible phytoalexins to combat fungal attack. The chemistry that elicits the formation of phytoalexins and the identification of small signalling molecules, such as methyl salicylate and methyl jasmonate, which induce the expression of silent gene clusters, has obvious importance in the context of plant protection. As the genomes of more fruits and vegetables are unravelled

Chemistry in the Kitchen Garden
By James R. Hanson
© James R. Hanson 2011
Published by the Royal Society of Chemistry, www.rsc.org

so their biosynthetic potential may become more apparent, giving guidance to those involved in plant breeding.

Many vegetables are cultivars of wild species and are grown in unnatural conditions of monoculture away from their country of origin. Examination of the genome of their wild relatives may reveal the potential for producing further valuable compounds for plant protection.

A major challenge to those who grow fruit and vegetables is the impact of climate change. A great deal has been written on its potential effect on the range of plants that may be grown in the garden. A number of hitherto exotic fruit and vegetables may become more easily grown, while a longer growing season may be of benefit to other plants. However, there is another side to the warmer winters. Microorganisms and insect herbivores are more likely to survive the winter and hence plant responses to these may need to be more effective earlier in the season. Younger leaves are often more vulnerable to attack. Thus the correlation that has been made between the mean January temperature, and the latitude at which cyanogenic glycosides are produced by clover, may be translated into the production of similar defensive substances by other plants. The warmer winters and hotter drier summers may also allow alien pests to flourish. The arrival of the harlequin ladybird (*Harmonia axyridis*) is a case in point. Indigenous leaf-miners such as the beet-leaf miner can cause serious damage to leaf crops, in this case, Swiss chard. Concern has been expressed about the spread of non-native leaf-miners. An example is a tomato leaf-miner (*Tuta absoluta*) which arrived from South America and has been spreading northwards from the Mediterranean. Its biocontrol by a parasitic wasp (*Dacnusa sibirica*) is possible in commercial greenhouses, whilst a pheromone trap based on (3*E*,8*Z*,11*Z*)-3,8,11-tetradecatrienyl acetate has also been used. There are many other examples of the continuing need to study the chemistry of phytophagous insect pests.

Another aspect of climate change involves the timing of flowering. Studies have revealed that earlier flowering is related to a rise in the average temperatures. In one study of orchids, flowering was found to be six days earlier for every degree increase in average spring temperatures. However, if a plant responds to climate change at a different rate from that of its pollinating insect, there could be a disruption to the pollination and fruiting sequence. The hormonal chemistry that underlies flowering may need investigation in this context. A number of plants require a cold period prior to flowering and this can be a problem. There is an impact of plant hormones on this aspect of plant development.

Any rise in temperature will affect the vapour pressure of volatile compounds, a feature revealed by a walk through a Mediterranean pine wood rather than through a wood in the UK. The ecological role of these plant volatiles as insect attractants and the distance over which they have an impact, is amplified by a change in temperature. As we have seen, the volatiles from many vegetables act as cues for phytophagous insects. The volatile compounds produced by plants as a consequence of damage by a herbivore can also act as a cue for the natural enemies of the herbivore. A target may be to breed cultivars which emit compounds that mediate these tripartite relationships.

A related effect is that of water loss. Long hot dry periods have an obvious impact on crop yields. There are campaigns to mitigate this by the use of domestic 'grey water'. Whilst this is a laudable use of a precious resource, it must be done with caution on foodstuffs. Grey water may include dissolved salts and bleaching agents as well as detergents that contain compounds of nitrogen and phosphorus which can damage leaf vegetables and alter the local mineral balance. Potentially harmful material including bacteria may adhere to leaf waxes, whilst food residues are insect attractants. It is therefore sensible to ensure that grey water is applied *via* the soil, where biodegradation and adsorption may take place. Rain water storage and the development of drought-tolerant vegetable cultivars is becoming a major way of mitigating the impact of climate change. The development of more deep-rooted cultivars and an understanding of the role of plant hormones, such as abscisic acid, in pore closure and response to water stress could be of benefit to crop production. The effect of climate change on soil temperature will clearly impact on the growth of microorganisms within the soil and on germination.

Part of the increase in crop yields over the last fifty years has been due to the utilization of chemicals for pest control. However, environmental and toxicity concerns, as well as the evolution of resistance to some pesticides, have led to the restriction in the use and the withdrawal of some pesticides. Understanding the chemistry of companion planting and the use of plant-based natural products may fill some of the gaps. Pesticides and herbicides based on natural product may have a shorter environmental half-life and, although they may be less efficient, they may have a more limited effect on the general wild life. For example non-selective weed management can be achieved with a 20% acetic acid solution or better with pelargonic acid (nonanoic acid) together with other organic acids such as lactic acid. The effect of the acidic environment of formicine ants can sometimes be seen on the grass around their nests. The allelopathic effects of constituents of plant essential oils, such as citral, citronellal and geraniol from lemon grass oil and citro-nella oil can be used. A number of the natural products obtained from rosemary or thyme can also have antimicrobial activity. Examination of gardening manuals from years gone by in the light of our modern knowledge of plant constituents, may give some useful leads. The chemistry of soil microorganisms can also provide acceptable plant protection agents. Thus bialaphos is a natural product that has herbi-cidal activity which is obtained from a soil Streptomycete (*S. hygro-scopis*) which is metabolized to phosphonathricin and which targets glutamine synthase and thus nitrogen uptake.

Another area in which there is a considerable potential chemical interest are the substances which mediate interactions between fruit and vegetables and their endophytic organisms, together with the role that these organisms play in the development of the plant. It has become apparent that almost all plants can harbour endophytic bacteria or fungi. In many cases there are beneficial symbiotic relationships. Endophytic bacteria and fungi have been isolated from cucumbers, peppers, tomatoes, pumpkins, potatoes, cabbages, lettuces, grapes and many other fruits. In some cases there is antagonism between the endophytic organism and phytopathogens. Thus a strain of *Fusarium oxysporum* EF119 from tomatoes showed activity against tomato late-blight (*Phytopthora infestans*), as well as damping-off organisms such as *Pythium ultimum*. A correlation has been observed between the presence of nonpathogenic *Fusarium* species in the root system, and resistance to soil-borne diseases and pests such as nematodes. Volatile antimicrobial compounds produced by an endophytic fungus, *Muscodor albus*, have been reported to provide some protection to potatoes against attack by nematodes. *Acremonium* and *Plectosporium* species found in lettuces had a beneficial effect on root growth. Sub-epidermal microorganisms have also been reported to give protection to fruit against post-harvest rot by, for example, *Rhizopus stolonifer*. A yeast-like fungus, *Aureobasidium pullulans* obtained from cherries, gave some biocontrol against post-harvest rot, whilst a *Pseudomonas* species from grapes inhibited the growth of *Botrytis cinerea*. Many of these and other mutualistic endo-phytic interactions may have a chemical basis and could provide lead compounds for plant protection. They may also affect the organoleptic properties of fruit and vegetables.

As we enter an era in which increasing demands will be placed on the production of food, the chemistry of fruit and vegetables will continue to provide a wide range of problems with potentially valuable results.

Further Readings

BOOKS

D. H. R. Barton and K. Nakanishi, *Comprehensive Natural Product Chemistry,* Elsevier, Amsterdam, 1999.

S. Berry and S. Bradley, *Plant Life, A Gardener's Guide,* Collins and Brown, London, 1993.

E. Block, *Garlic and Other Alliums, The Lore and The Science,* The Royal Society of Chemistry, Cambridge, 2010.

J. Buckingham, *Dictionary of Natural Products,* Chapman and Hall, London, 1994.

T. P. Coultate, *Food, The Chemistry and its Components,* The Royal Society of Chemistry, Cambridge, 4th edn, 2002.

R. J. Cremlyn, *Agrochemicals, Preparation and Mode of Action*, John Wiley, Chichester, 1991.

M. Cresser, K. Killham and T. Edwards, *Soil Chemistry and its Applications,* Cambridge University Press, Cambridge, 1993.

P. M. Dewick, *Medicinal Natural Products – A Biosynthetic Approach,* John Wiley, Chichester, 2nd edn. 2000.

R. Finn, *Nature's Chemicals, The Natural Products that Shaped our World*, Oxford University Press, Oxford, 2010.

J. R. Hanson, *Natural Products, The Secondary Metabolites,* The Royal Society of Chemistry, Cambridge, 2003.

J. R. Hanson, *Chemistry in the Garden,* The Royal Society of Chemistry, Cambridge, 2007.

J. R. Hanson, *The Chemistry of Fungi,* The Royal Society of Chemistry, Cambridge, 2008.

Chemistry in the Kitchen Garden
By James R. Hanson
© James R. Hanson 2011
Published by the Royal Society of Chemistry, www.rsc.org

J. B. Harborne, *Introduction to Ecological Biochemistry,* Academic Press, London 4th edn, 1993.

J. B. Harborne and H. Baxter, *Chemical Dictionary of Economic Plants,* John Wiley, Chichester, 1997.

J. B. Harborne, H. Baxter and G. P. Moss, *Dictionary of Plant Toxins,* John Wiley, Chichester, 1997.

D. S. Ingram, D. Vince-Prue and P. J. Gregory, *Science and the Garden,* Royal Horticultural Society and Blackwells, Oxford, 2002.

J. Mann, *Chemical Aspects of Biosynthesis,* Oxford University Press, Oxford, 1992.

J. Mann, *Murder, Magic and Medicine,* Oxford University Press, Oxford, 1992.

I. Ridge, *Plants,* The Open University and Oxford University Press, Oxford, 2002.

C. S. Sell, *A Fragrant Introduction to Terpenoid Chemistry,* The Royal Society of Chemistry, Cambridge, 2003.

W. Steglich, B. Fugmann and S. Lang-Fugmann, *ROMPP Encyclopedia of Natural Products,* Thieme, Stuttgart, 2000.

R. H. Thomson, *The Chemistry of Natural Products*, Blackie, London, 2nd edn, 1993.

J. Wong, *Grow Your Own Drugs,* Collins, London, 2009.

REVIEWS

The review journal, *Natural Product Reports*, published by The Royal Society of Chemistry contains articles covering particular groups of natural products and their biological activity which are relevant to the chemistry of fruit and vegetables. Journals which contain articles and reviews of particular interest include *the Journal of Agricultural and Food Chemistry*, *Food and Function*, *The Journal of Chemical Ecology*, *The Journal of Natural Products*, and *Phytochemistry*.

SPECIFIC ARTICLES AND REVIEWS

Chapter 1

M. H. Beale and J. L.Ward, Jasmonates, key players in the plant defence, *Nat. Prod. Rep.,* 1998, **15**, 533–548.

J. Burns, P. D. Fraser and P. M. Bramley, Identification and quantification of carotenoids, tocopherols and chlorophylls in commonly consumed fruit and vegetables, *Phytochemistry,* 2003, **62**, 939–947.

A. Crozier, I. B. Jaganath and M. N. Clifford, Dietary phenolics: chemistry, bioavailability and effects on health, *Nat. Prod. Rep.,* 2009, **26**, 1001–1043.

R. H. J. Erkens, What every chemist should know about plant names. *Nat. Prod. Rep.*, 2011, **28**, 11–14.

S. Fujioka and A. Sakurai, Brassinosteroids, *Nat. Prod. Rep.,* 1997, **14**, 1–10.

A. R. Knaggs, Biosynthesis of shikimate metabolites, *Nat. Prod. Rep.,* 2003, **20**, 119–136.

T. Kuzuyama and H. Seto, Diversity of the biosynthesis of the isoprene units, *Nat. Prod. Rep.*, 2003, **20**, 171–183.

J. MacMillan, The biosynthesis of the gibberellin plant hormones, *Nat. Prod. Rep.,* 1997, **14**, 221–243.

L. N. Mander, Twenty years of gibberellin research, *Nat. Prod. Rep.,* 2003, **20**, 49–69.

T. Oritani and H. Kiyota, Biosynthesis and metabolism of abscisic acid and related compounds, *Nat. Prod. Rep.,* 2003, **20**, 414–425.

M.-H. Pan, C.-S. Lai and C.-T. Ho, Anti-inflammatory activity of natural dietary flavonoids, *Food Funct.,* 2010, **1**, 15–31.

F. Poiroux-Gonord, L. P. R. Bidel, A. L. Fanciullino, H. Gautier, F. Lauri-Lopez and L. Urban, Health benefits of vitamins and secondary metabolites of fruit and vegetables and prospects to increase their concentrations by agronomic approaches, *J. Agric. Food Chem.*, 2010, **58**, 12065–12082.

K. Springob, T. Nakajima, M. Yamazaki and K. Saito, Recent advances in the biosynthesis of anthocyanins, *Nat. Prod. Rep.,* 2003, **20**, 288–303.

M. Zagrobelny, S. Bak, A. V. Raasmussen, B. Jorgensen, C. M. Naumann and B. L. Moller, Cyanogenic glucosides and plant-insect interactions, *Phytochemistry,* 2004, **65**, 293–306.

Chapter 2

M. A. Birkett, C. A. M. Campbell, K. Chamberlain, E. Guerrieri, A. J. Hick, J. L. Martin, M. Matthes, J. A. Napier, J. Pettersson, J. A. Pickett, G. M. Poppy, E. M. Pow, B. J. Pye, L. E. Smart, G. H. Wadhams, L. J. Wadhams and C. M. Woodcock, New roles for cis-jasmone as an insect semiochemical and in plant defense, *Proc. Nat. Acad. Sci. U. S. A.,* 2000, **97**, 9329–9334.

J. Engelberth, H. T. Alborn, E. A. Schmelz and J. H. Tumlinson, Airborne signals prime plants against insect herbivore attack, *Proc. Nat. Acad. Sci. U. S. A.,* 2004, **101**, 1781–1785.

S. K. Gaw, N. D. Kim, G. L. Northcott, A. L. Wilkins and G. Robinson, Uptake of DDT, arsenic, cadmium, copper and lead by lettuce and radish grown in contaminated horticultural soils, *J. Agric. Food Chem.*, 2008, **56**, 6584–6593.

M. E. Lights, B. V. Burger, L. Kohout and J. van Staden, Butenolides from plant-derived smoke: natural plant growth regulators with antagonistic actions on seed germination, *J. Nat. Prod.*, 2010, **73**, 267–269.

O. Lopez, J. G. Fernandez-Bolanos and M. V. Gill, New trends in pest control: the search for greener insecticides, *Green Chem.*, 2005, **7**, 431–442.

R. J. Petroski and D. W. Stanley, Natural compounds for pest and weed control, *J. Agric. Food Chem.*, 2009, **57**, 8171–8179.

A. E. Smith and D. M. Secoy, Plants used for agricutural pest control in Western Europe before 1850, *Chem. Ind.*, 1981, 12–17.

J. R. Vyvyan, Allelochemicals and leads for new herbicides and agrochemicals, *Tetrahedron,* 2002, **58**, 1631–1646.

K. Wenke, M. Kai and B. Piechulla, Below ground volatiles facilitate interactions between plant roots and soil organisms, *Planta,* 2010, **231**, 499–506.

Chapter 3

C. Alasavar, J. M.Grigor, D. Zhang, P. C. Quantick and F. Shahidi, Comparison of volatiles, phenolics, sugars, anti-oxidant vitamins and sensory qualities of different coloured carrot varieties, *J. Agric. Food Chem.*, 2001, **49**, 1410–1416.

E. Block, The organosuifur chemistry of the genus, *Allium, Angew. Chem., Int. Ed.,* 1992, **31**, 1138–1178.

E. Block, A. J. Dane, S. Thomas and R. B. Cody, Applications of direct analysis in real time mass spectrometry (DART-MS) in *Allium* chemistry, *J. Agric. Food Chem.*, 2010, **58**, 4617–4625.

A. Czepa and T. Hofmann, Structural and sensory characterization of compounds contributing to the bitter off-taste of carrots, *J. Agric. Food Chem.*, 2003, **51**, 3865–3873.

M. Friedman, Chemistry, biochemistry and dietary role of potato polyphenols, *J. Agric. Food Chem.*, 1997, **45**, 1523–1540.

M. Friedman, Potato glycoalkaloids and metabolites: roles in the plant, *J. Agric. Food Chem.*, 2006, **54**, 8655–8681.

X. Huang and L. Kong, Steroidal saponins from the roots of *Asparagus officinalis, Steroids,* 2006, **71**, 171–176.

M. Imsic, S. Winkler, B. Tomkins and R. James, Effect of storage and cooking on β-carotene isomers in carrots, *J. Agric, Food Chem.,* 2010, **58**, 5109–5113.

Y. Ito, N. Sugimoto, T. Akiyama, T. Yamazaki and K. Tanamoto, Cepaic acid, a novel yellow xanthylium pigment from the dried outer scales of the yellow onion, *Allium cepa, Tetrahedron Lett.,* 2009, **50**, 4084–4086.

Y. Koda, Y. Kikuta, H. Tazaki, Y. Tsujino, S. Sakamura and T. Yoshihara, Potato tuber inducing activities of jasmonic acid and related compounds, *Phytochemistry,* 1991, **30**, 1435–1440.

M. Petersen and M. J. Simmonds, Rosmarinic acid, *Phytochemistry,* 2003, **62**, 121–125.

S. Purup, E. Larsen and L. P. Christensen, Differential effects of falcarinol and related C_{17} polyacetylenes on intestinal cell proliferation, *J. Agric. Food Chem.,* 2009, **57**, 8290–8296.

R. Slimestad, T. Fossen and I. M. Vagen, Onions: a source of unique dietary flavonoids, *J. Agric. Food Chem.,* 2007, **55**, 10067–10080.

D. Strack, T. Vogt and W. Schliemann, Recent advances in betalain research, *Phytochemistry,* 2003, **62**, 247–269.

P. Thipyapong, M. D. Hunt and J. C. Steffens, Systemic wound induction in potato polyphenol oxidase, *Phytochemistry,* 1995, **40**, 673–676.

C. Zidorn, K. Johrer, M. Ganzera, B. Schubert, E. M. Sigmund, J. Mader, R. Greil, E. P. Ellmerer and H. Stuppner, Polyacetylenes from the *Apiaceae* vegetables, carrot, celery, fennel, parsley and parsnip and their cytotoxic activities. *J. Agric. Food Chem.,* 2005, **53**, 2518–2523.

Chapter 4

A. Blaakmeer, A. Stork, A. van Veldhuizen, T. A. van Beek, A. de Groot, J. A. van Loon and L. M. Schoonhoven, Isolation, identification and synthesis of miriamides, new host markers from eggs of *Pieris brassicae, J. Nat. Prod.,* 1994, **57**, 90–99.

P. D. Brown and M. J. Morra, *Brassicaceae* tissues as inhibitors of nitrification in soil, *J. Agric. Food Chem.,* 2009, **57**, 7706–7711.

R. Edenharder, G. Keller, K. L. Platt and K. K. Uinger, Isolation and characterization of structurally novel anti-mutagenic flavonoids from spinach (*Spinacia oleracea*), *J. Agric. Food Chem.,* 2001, **49**, 2767–2773.

M. H. Gordon, Dietary antioxidants in disease prevention, *Nat. Prod. Rep.,* 1996, **13**, 265–273.

R. Holst and G. Williamson, A critical review of the bioavailability of glucosinolates, *Nat. Prod. Rep.*, 2004, **21**, 425–447.

M. Bjorkamn, I. Klingen, A. N. E. Birch, A. M. Bones, T. J. A. Bruce, T. J. Johansen, R. Meadow, J. Molmann, R. Seljasen, L. E. Smart and D. Stewart, Phytochemicals of Brassicaceae in plant protection and human health – influences of climate, environment and agronomic practice, *Phytochemistry*, 2011, **72**, 538–556.

R. Kronbak, F. Duus and O. Vang, Effect of 4-methoxyindole-3-carbinol on the proliferation of colon cancer cells in vitro, *J. Agric. Food Chem.*, 2010, **58**, 8453–8459.

J. Kruk, Occurrence of chlorophyll precursors in leaves of cabbage heads, the case of natual etiolation, *Photochem. Photobiol.*, 2005, **80**, 187–194.

H. T. Le, C. M. Schaldach, G. L. Firestone and L. F. Bjeldanes, Plant derived 3,3′-diindolylmethane is a strong androgen antagonist in human prostate cancer cells, *J. Biol. Chem.*, 2003, **278**, 21136–21148.

R. Lo Scalzo, A. Genna, F. Branca, M. Chedin and H. Chassaigne, Anthocyanin composition of cauliflower and cabbage, *Food Chem.*, 2008, **107**, 136–144.

J.-K. Moon, J.-R. Kim, Y.-J. Ahn and T. Shibamoto, Analysis and anti-Helicobacter activity of sulforaphane and related compounds present in broccoli, *J. Agric. Food Chem.*, 2010, **58**, 6672–6677.

R. F. Mithen, B. G. Lewis, R. K. Heaney and G. R. Fenwick, Glucosinolates of wild and cultivated Brassica species, *Phytochemistry*, 1987, **26**, 1969–1973.

S. Mukherjee, A. Gangopadhyay and D. K. Das, Broccoli: A unique vegetable that protects mammalian hearts through the redox cycling of the thioredoxin family, *J. Agric. Food Chem.*, 2008, **56**, 609–617.

M. S. C. Pedras, F. I. Okanga, I. L. Zaharia and A. Q. Khan, Phytoalexins from crucifers: Synthesis, biosynthesis and biotransformation, *Phytochemistry*, 2000, **53**, 161–176.

M. S. C. Pedras and E. E. Yaya, Phytoalexins from *Brassicaceae* news from the front, *Phytochemistry*, 2010, **71**, 1191–1197.

R. A. Sessa, M. H. Bennett, M. J. Lewis, J. W. Mansfield and M. H. Beale, Metabolite profiling of sesquiterpene lactones from *Lactuca* species, *J. Biol. Chem.*, 2000, **275**, 26877–26884.

L. G. West, K. A. Meyer, B. A. Balch, F. J. Rosse, M. R. Schultz and G. W. Haas, Glucoraphanin and 4-hydroxyglucobrassicin contents of 59 cultivars of broccoli, raab, kohlrabi, radish, cauliflower, kale and cabbage, *J. Agric. Food Chem.*, 2004, **52**, 916–926.

Chapter 5

C. Bomke and B. Tudzynski, Diversity, regulation and evolution of the gibberellin biosynthetic pathway in fungi compared to plants and bacteria, *Phytochemistry,* 2009, **70,** 1876–1893.

T. Bruce, M. Birkett, J. Blande, A. M. Hooper, J. L. Martin, B. Khambay, I. Prosser, L. E. Smart and L. J. Wadhams, Response of economically important aphids to components of *Hemizygia petiolata* essential oil, *Pest Manag. Sci.,* 2005, **61,** 1115–1121.

T. Cornwell, W. Cohick and I. Raskin, Dietary phytoestrogens and health, *Phytochemistry,* 2004, **65,** 995–1016.

V. M. Sponsel, P. Gaskin and J. MacMillan, Identification of gibberellins in immature seeds of *Vicia faba* and some chemotaxonomic considerations, *Planta,* 1979, **146,** 101–105.

B. Webster, T. Bruce, S. Dufour, C. Birkemeyer, M. Birkett, J. Hardie and J. Pickett, Identification of volatile compounds used in host location by the black bean aphid, *J. Chem. Ecol.,* 2008, **34,** 1153–1161.

X. Xie, K. Yoneyama, Y. Harada, N. Fusegi, Y. Yamada, S. Ito, T. Yokota, Y. Takeuchi and K. Yoneyama, Fabacyl acetate, a germination stimulant for root parasitic plants from *Pisum sativum, Phytochemistry,* 2009, **70,** 211–215.

Chapter 6

C. H. Azevedo-Meleiro and D. B. Rodriguez-Amaya, Qualitative and quantitative differences in carotenoid composition among *Cucurbita moschata, C.maxima* and *C.pepo, J. Agric. Food Chem.,* 2007, **55,** 4027–4033.

S. Birtic, C. Ginles, M. Causse, C. M. G. C. Renard and D. Page, Changes in volatiles and glycosides during fruit maturation of two contrasted tomato (*Solanum lycopersicon*) lines, *J. Agric. Food Chem.,* 2009, **57,** 591–598.

S. Blechschmidt, U. Castel, P. Gaskin, P. Hedden, J. E. Graebe and J. MacMillan, GC/MS analysis of the plant hormones in seeds of *Cucurbita maxima, Phytochemistry,* 1984, **23,** 553–558.

B. H. Davies, S. Matthews and J. T. O. Kirk, Nature and biosynthesis of the carotenoids of different colour varieties of *Capsicum annuum, Phytochemistry,* 1970, **9,** 797–805.

J. M. Grunzweig, H. D. Rabinowitch, J. Katan, M. Wodner and Y. Ben-Tal, Endogenous gibberellins in foliage of tomato, *Phytochemistry,* 1997, **46,** 811–815.

D. Hervert-Hernandez, S. G. Sayago-Ayerdi and I. Gani, Bioactive compounds of four hot pepper varieties (*Capsicum annuum*), antioxidant capacity and intestinal bioaccessibility, *J. Agric. Food Chem.*, 2010, **58**, 3309–3406.

R. E. Kopec, K. M. Riedl, E. H. Harrison, R. W. Curley Jr., D. P. Hruszkewycz, S. K. Clinton and S. J. Schwartz, Identification and quantification of apo-lycopenals in fruits, vegetables and human plasma, *J. Agric. Food Chem.*, 2010, **58**, 3290–3296.

P. Ortiz-Serrano and J. V. Gil, Quantitative comparison of free and bound volatiles of two commercial tomato cultivars (*Solanum lycopersicum*) during ripening, *J. Agric. Food Chem.*, 2010, **58**, 1106–1114.

J. C. Serrani, B. Sanjuan, O. Ruiz-Rivero, M. Fos and J. L. Garcia-Martinez, Gibberellin regulation of fruit set and growth in tomato, *Plant Physiol.*, 2007, **145**, 246–257.

S. Yahara, N. Uda, E. Yoshio and E. Yae, Steroidal alkaloid glycosides from tomato, *J. Nat. Prod.*, 2004, **67**, 500–502.

Chapter 7

K. Aaby, G. Skrede and R. Wrolstad, Phenolic composition and antioxidant activities in flesh and achenes of strawberries, *J. Agric. Food Chem.*, 2005, **53**, 4032–4040.

P. S. Blake, D. R. Taylor, C. M. Crisp, L. N. Mander and D. J. Owen, Identification of endogenous gibberellins in strawberry including the novel gibberellins, GA_{123}, GA_{124} and GA_{125}, *Phytochemistry*, 2000, **55**, 887–890.

G. Borges, A. Degeneve, W. Mullen and A. Crozier, Identification of flavonoid and phenolic anti-oxidants in blackcurrants, blueberries, raspberries, redcurrants and cranberries, *J. Agric. Food Chem.*, 2010, **58**, 3901–3909.

E. Gomez, C. A. Ledbetter and P. L. Hartsell, Volatile compounds in apricot, plum and their interspecies hybrids, *J. Agric. Food Chem.*, 1993, **41**, 1669–1676.

X. He and R. H. Liu, Triterpenoids isolated from apple peels have potent anti-proliferative activity and may be partially responsible for apple's anti-cancer activity, *J. Agric. Food Chem.*, 2007, **55**, 4366–4370.

R. Krikorian, M. D. Shidler, T. A. Nash, W. Kalt, M. R. Vinqvist-Tymchuk, B. Shukitt-Hale and J. A. Joseph, Blueberry supplementation improves memory in older adults. *J. Agric. Food Chem.*, 2010, **58**, 3996–4000.

S. C. Marks, W. Mullen and A. Crozier, Flavonoid and hydro-xycinnamate profiles of English apple cider, *J. Agric. Food Chem.*, 2007, **55**, 8723–8730.

J. H. J. Martin, S. Crotty, P. Warren and P. N. Nelson, Does an apple a day keep the doctor away because a phytoestrogen a day keeps the virus at bay? A review of the anti-viral properties of phytoestrogens, *Phytochemistry*, 2007, **68**, 266–274.

E. Mehangic, G. Royer, R. Symoneaux, F. Jourjon and C. Prost, Characterization of odor-active volatiles in apples: influence of culti-vars and maturity stage, *J. Agric. Food Chem.*, 2006, **54**, 2678–2687.

A. V. Rao and D. M. Snyder, Raspberries and human health: A review, *J. Agric. Food Chem.*, 2010, **58**, 3871–3883.

B. Roschek, R. C. Fink, M. D. McMichael, D. Li and R. S. Alberte, Elderberry flavonoids bind to and prevent H1N1 infection *in vitro*, *Phytochemistry*, 2009, **70**, 1255–1261.

B. Schwarz and T. Hofmann, Sensory-guided decomposition of red-currant juice (*Ribes rubrum*) and structure determination of key astringent compounds, *J. Agric. Food Chem.*, 2007, **55**, 1394–1404.

N. P. Seeram, L. S. Adams, Y. Zhang, R. Lee, D. Sand, H. S. Scheuller and D. Heber, Blackberry, black raspberry, blueberry, cranberry, red raspberry and strawberry extracts inhibit growth and stimulate apoptosis of human cancer cells *in vitro*, *J. Agric. Food Chem.*, 2006, **54**, 9329–9339.

M. Singh, M. Arseneault, T. Sanderson, V. Murthy and C. Ramassamy, Challenges for research on polyphenols from foods in Alzheimer's disease: bioavailability, metabolism, and cellular and molecular mechanisms, *J. Agric. Food Chem.*, 2008, **56**, 4855–4873.

A. Solomon, S. Galubowicz, Y. Yablowicz, S. Grossman, M. Bergman, H. E. Gottlieb, A. Altman, Z. Kerem and M. A. Flaishman, Anti-oxidant activities and anthocyanin content of fresh fruits of common fig (*Ficus carica*), *J. Agric. Food Chem.*, 2006, **54**, 7717–7723.

P. Vanzani, M. Rossetto, A. Rigo, U. Vrhovsek, F. Mattivi, E. D'Amato and M. Scarpa, Major phytochemicals in apple cultivars: contribution to peroxyl radical trapping efficiency, *J. Agric. Food Chem.*, 2005, **53**, 3377–3382.

K. Wolfe, X. Wu and R. H. Liu, Anti-oxidant activity of apple peels, *J. Agric. Food Chem.*, 2003, **51**, 609–614.

K. Wolfe and R. H. Liu, Apple peels as a value-added food ingredient, *J. Agric. Food. Chem.*, 2003, **51**, 1676–1683.

W. Yi, J. Fischer, G. Krewer and C. C. Akoh, Phenolic compounds from blueberries can inhibit colon cancer ceil proliferation and induce apoptosis, *J. Agric. Food Chem.*, 2005, **53**, 7320–7329.

R. Zadernowski, M. Naczk and J. Nesterowicz, Phenolic acid profiles in some small berries, *J. Agric. Food Chem.,* 2005, **53**, 2118–2124.

Y. Zhang, N. P. Seerman, R. Lee, L. Feng and D. Heber, Isolation and identification of strawberry phenolics with anti-oxidant and human cancer cell anti-proliferative properties, *J. Agric. Food Chem.,* 2008, **56**, 670–675.

Chapter 8

L. G. Angelini, G. Carpanese, P. L. Cieni, I. Morelli, M. Macchia and G. Flamini, Essential oils from Mediterranean *Lamiaceae* as weed germination inhibitors, *J. Agric. Food Chem.,* 2003, **51**, 6158–6164.

N. Bai, K. He, M. Roller, C.-S. Li, X. Shao, M.-H. Pan and C.-T. Ho, Flavonoids and phenolic compounds from *Rosmarinus officinalis, J. Agric. Food Chem.,* 2010, **58**, 5363–5367.

H. J. Bouwmeester, J. A. R. Davies and H. Toxopeus, Enantiomeric composition of carvone, limonene and carveols in seeds of dill and annual and biennial caraway varieties, *J. Agric. Food Chem.,* 1995, **43**, 3057–3064.

L. B. Buck, Unraveling the sense of smell, *Angew. Chem., Int. Ed.,* 2005, **44**, 6128–6140.

P. L. Cantore, N. S. Iacobellis, A. de Marco, F. Capasso and F. Senatore, Anti-bacterial activity of *Coriandrum sativum* and *Foeniculum vulgare* essential oils, *J. Agric. Food Chem.,* 2004, **52**, 7862–7866.

A. Careeda, B. Marongiu, S. Poriedda and C. Soro, Supercritical carbon dioxide extraction and characterization of *Laurus nobilis* essential oil, *J. Agric. Food Chem.,* 2002, **50**, 1492–1496.

I. Dadalioglu and G. A. Evrendilek, Chemical compositions and anti-bacterial effects of essential oils of Turkish oregano, (*Origanum minutiflorum*), bay laurel (*Laurus nobilis*), Spanish lavender (*Lavandula stoechas*) and fennel (*Foeniculum vulgare*) on common foodborne pathogens, *J. Agric. Food Chem.,* 2004, **52**, 8255–8260.

H. J. Dorman, M. Kosar, K. Kahlos, Y. Holm and R. Hiltunen, Anti-oxidant properties and composition of aqueous extracts from *Mentha* species, hybrids, varieties and cultivars, *J. Agric. Food Chem.,* 2003, **51**, 4563–4569.

G. Franzios, M. Mirotsou, E. Hatziapostolou, J. Kral, Z., G. Scouras and P. Mavragani-Tsipidou, Insecticidal and genotoxic activities of mint essential oils, *J. Agric. Food Chem.,* 1997, **45**, 2690–2694.

H. G. Maier, Volatile flavouring substances in foodstuffs, *Angew. Chem., Int. Ed.,* 1970, **9**, 917–988.

M. W. Perlino, C. Theoduloz, J. A. Rodriguez, T. Yanez, V. Lazo and G. Schmeda-Hirschmann, Gastroprotective effect of carnosic acid γ-lactone, *J. Nat. Prod.*, 2010, **73**, 639–643.

J. E. Simon and J. Quinn, Characterization of essential oil of parsley, *J. Agric. Food Chem.*, 1988, **36**, 467–472.

The Families of Common Fruit and Vegetables

The formation of certain classes of natural product can be a characteristic of a particular family of plants. Even though different parts of a plant may be used for a food, when comparing the constituents of two members of the same family, the presence of these characteristic natural products may be revealed. Consequently, it can be helpful to identify the plant family to which a fruit or vegetable belongs. The following list of plant families shows the common fruit and vegetables which they include.

Alliaceae (once part of the Liliaceae)
Onions	*Allium cepa*
Garlic	*Allium sativum*
Leeks	*Allium ampeloprasum* var. *porrum*
Chives	*Allium schoenoprasum*

Amaranthaceae (Chenopodiaceae)
Beetroot	*Beta vulgaris*
Spinach	*Spinacia oleracea*

Apiaceae (Umbelliferae)
Carrots	*Daucus carota*
Parsnip	*Pastinaca sativa*
Celery	*Apium graveolens*

Chemistry in the Kitchen Garden
By James R. Hanson
© James R. Hanson 2011
Published by the Royal Society of Chemistry, www.rsc.org

Parsley   *Petroselinum crispum*
Coriander   *Coriandrum sativum*
Dill   *Anethum graveolens*
Fennel   *Foeniculum vulgare*

Asparagaceae (once part of the Liliaceae)
Asparagus   *Asparagus officinalis*

Asteraceae (Compositae)
Lettuce   *Lactuca sativa*
Globe artichoke   *Cynara cardunculus*
Chicory   *Chicorium intybus*
Endive   *Cichorium endivia*
Curry plant   *Helichrysum italicum* or *H. angustifolium*

Brassicaceae (Cruciferae)
Cabbage   *Brassica oleracea* (Capitata group)
Brussel sprouts   *Brassica oleracea* (Gemmifera group)
Broccoli   *Brassica oleracea* (Italica group)
Cauliflower   *Brassica oleracea* (Botrytis group)
Kale or Spring greens   *Brassica oleracea* (Acephala group)
Radish   *Raphanus sativus*
Black mustard   *Brassica nigra*
White mustard   *Brassica alba*
Swede   *Brassica napus*
Turnip   *Brassica rapa*

Cucurbitaceae
Cucumber   *Cucumis sativus*
Pumpkin   *Cucurbita maxima*
Marrow   *Cucurbita moschata*
Courgette   *Cucurbita pepo*

Fabaceae (Leguminosae)
Peas   *Pisum sativum*
Broad bean   *Vicia faba*
Runner bean   *Phaseolus coccineus*
French bean   *Phaseolus vulgaris*

Grossulariaceae
Blackcurrant	*Ribes nigrum*
Redcurrant	*Ribes rubrum*
Gooseberry	*Ribes uva-crispa*

Lamiaceae (Labiatae)
Mints
Spearmint	*Mentha spicata*
Cornmint	*Mentha arvensis*
Pennyroyal	*Mentha pulegium*
Peppermint	*Mentha x piperita*
Applemint	*Mentha rotundifolia*
Basil	*Ocimum basilicum*
Thyme	*Thymus vulgaris*
Lemon balm	*Melissa officinalis*
Rosemary	*Rosmarinus officinalis*
Sage	*Salvia officinalis*
Hyssop	*Hyssop officinalis*

Lauraceae
Bay laurel	*Laurus nobilis*

Moraceae
Fig	*Ficus carica*

Poaceae (Gramineae)
Sweet corn (maize)	*Zea mays*

Polygonaceae
Rhubarb	*Rheum rhabarbarum* or *Rheum x hybridum*

Rosaceae
Apple	*Malus domestics*
Pear	*Pyrus communis*
Quince	*Cydonia oblonga*
Plum	*Prunus domestica*
Cherry	*Prunus avium*
Peach	*Prunus persica*

Raspberry	*Rubus idaeus*
Blackberry	*Rubus fruticosus*
Strawberry	*Fragaria x ananassa*

Solanaceae

Potato	*Solanum tuberosum*
Tomato	*Solanum lycopersicum*
Aubergine	*Solanum melongena*
Peppers	*Capsicum annuum*
	(some are *Piper nigrum*, family Piperaceae)

Vitaceae

| Grapes | *Vitis vinifera* |

Glossary

Abscission: the loss of leaves, flowers and fruit from plants. The process is mediated by the hormone, abscisic acid.

Achene: a small dry fruit which does not open to release the seed. The 'seeds' on the surface of the strawberry are achenes. The edible fleshy part of the strawberry, commonly called the fruit, is an accessory tissue.

Adventitious: shoots and roots which appear from an unusual part of a plant when, for example, a cutting is rooted.

Aglycone: the non-sugar portion, often a terpenoid, alkaloid or phenylpropanoid, of a glycoside (*q.v.*)

Algae: simple photosynthetic plants which are not divided into roots, stems and leaves.

Alkaloid: a large group of structurally diverse, basic nitrogenous natural products.

Allelopathy: the release of a compound by an organism which inhibits the growth of other organisms in the locality.

Alliaceae: a family of plants which includes the Alliums (garlic and onions).

Allomone: a compound which is produced by one organism that has a detrimental allelopathic effect on a member of another species.

Amylase: an enzyme which catalyzes the hydrolysis of starch.

Anther: the upper part of the stamen which contains the pollen.

Chemistry in the Kitchen Garden
By James R. Hanson
© James R. Hanson 2011
Published by the Royal Society of Chemistry, www.rsc.org

Anthocyanin: a family of oxygen heterocycles that are biosynthesized by a combination of the phenylpropanoid and polyketide pathways and which contribute to the colours of plants.

Apiaceae: a family of plants, formerly known as the Umbelliferae and which include the carrots and celery.

Apical dominance: a situation in which the bud at the apex of the stem prevents the development of lateral branches. Removal of the bud allows the lateral branches to develop.

Asteraceae: a family of plants formerly known as the Compositae or daisy family and which includes the lettuce.

Auxin: a plant growth hormone such as indolylacetic acid, which promotes cell elongation rather than cell division.

Bacteria: ubiquitous microscopic unicellular organisms which are prokaryotes (lacking a true nucleus) and which have important roles in the decay of organic matter and in the fixation of nitrogen.

Betalains: a family of nitrogenous pigments which are iminium salts of betalamic acid, a tetrahydropyridine-2,6-dicarboxylic acid. They are found in beetroots.

Bioremediation: the use of biological means to restore contaminated land.

Biosynthesis: the biological synthesis of a natural product. The term 'biotransformation' usually refers to the biological conversion of a substance that is alien to the transforming organism.

Brassicaceae: a family of plants formerly known as the Cruciferae, which include the cabbages.

Brassinosteroids: a group of steroidal plant hormones which affect the response of a plant to stress.

C-3 and C-4 plants: plants in which the first products of photosynthesis contain either three or four carbon atoms.

Callus tissue: undifferentiated tissue which is often found growing over a wound.

Calyx: the collective term for the outer leaf-like part (sepal) of a flower.

Cambium: a layer of cells between the xylem and the phloem.

Carabid beetle: ground beetle that is often brown or black and is found under wood, stones or leaf letter.

Carbohydrate: a generic name for sugars including mono-, di- and higher saccharides.

Carotenoid: a group of mainly C_{40} isoprenoid pigments with a conjugated polyene chromophore and which includes β-carotene, a pigment of carrots.

Cellulose: a polysaccharide composed of chains of glucose units and which forms the plant cell wall.

Chemosystematics: the use of the natural products formed by plants as part of the basis for their classification.

Chloroplast: a specialized structure within the cells of plants that contain chlorophyll and are the site of photosynthesis.

Climacteric: a period of increased respiration in the ripening of some fruit often post-harvest, which is stimulated by ethylene gas. Fruit such as apples, bananas and tomatoes are climacteric whilst others like grapes and strawberries, are not.

Cork: a layer of protective tissue that is found in woody plants just below the epidermis.

Cotyledon: a specialized leaf-like structure, which is often swollen with food reserves and which makes up part of the embryo in a seed. A plant which is a monocotyledon has one and a dicotyledon has two cotyledons in the seed.

Cucurbitaceae: a plant family with a typical vine-like growth which includes the cucumbers, squashes, marrows and melons. They are sometimes known as the 'cucurbits'.

Cultivar: a variety of a plant which has been raised in cultivation rather than being obtained from the wild.

Cytokinins: a group of plant hormones which are formed in the roots and which promote cell division.

Dormancy: a resting period in plant growth and development when there is a reduced metabolic rate.

Ecdysis: the moulting or shedding of the cuticle in insects which is usually accompanied by a significant change in size or form. The process is regulated by the presence of the juvenile moulting hormone and the ecdysteroids.

Electroantennogram: when the insect antenna receptors are stimulated by interaction with a semiochemical, a potential difference is created between the tip and the base. This difference is recorded as the electroantennogram and is a measure of the receptors that are stimulated.

Endophyte: a microorganism growing within a plant.

Etiolated: a condition in which the growing plant is weakened and is producing relatively little chlorophyll.

Fabaceae: the pea and bean family of plants formerly known as the Leguminosae.

Feeder roots: fine branch roots that are involved in the uptake of nutrients.

Flavonoid: a group of oxygen heterocycles which are biosynthesized by a mixed phenylpropanoid/polyketide pathway and include many plant pigments such as the flavones and anthocyanidins. They are often polyphenols and act as powerful antioxidants.

Formicine ants: ants which produce formic acid.

Fungi: eukaryotic microorganisms which are non-photosynthetic and which obtain their nutrients by degradation and by absorption from their surroundings.

Genus: in the taxonomic classification of plants, the level that comes between species and family. In the binomial plant name, the genus comes first. Plants of the same genus have similar morphological features.

Gibberellins: a group of plant hormones that regulate several aspects of plant growth and development including germination and cell elongation.

Glycoside: a composite molecule comprising one or more sugars attached to an aglycone such as a terpenoid, alkaloid, phenylpropanoid or polyketide.

Heartwood: the wood which forms the central region of a tree trunk. It often contains resin and provides mechanical support to the tree.

Hyphae: microscopic thread-like structures which make up the living fungus.

Hypocotyl: in a seedling it is the region of the axis between the cotyledons and the roots.

Kairomone: a group of pheromones which are interspecies specific and benefit the receiver. Thus when a pine tree is damaged by a beetle, it may produce a terpene such as myrcene which lures more beetles to the tree. The myrcene is a kairomone and the beetles are the beneficiaries.

Lamiaceae: a family of plants, formerly known as the Labiatae which include the mints, rosemary, sage and other culinary herbs.

Latex: a gummy liquid exudate from a fungus or a plant.

Lichen: a composite organism which is formed by the symbiotic association between a fungus and an alga or cyanobacterium.

Lignin: a polymer formed from aromatic C_6-C_3 units which provide the strengthening material of plants.

Lipid: a hydrophobic, fat-soluble natural product. They include the fatty acids, plant waxes, plant sterols and glyceryl esters.

Metabolites: the generic name for compounds formed by living systems. Primary metabolites are compounds which occur in all living cells and play a central role in the metabolism and reproduction of the cell. Secondary metabolites are characteristic of a limited range of species and many exert their biological effects on other cells or organisms.

Mitochondria: a sub-cellular structure (organelle) in which respiration occurs.

Mycelium: the mat of growing fungal hyphae.

Mycorrhizal fungi: fungi which have a symbiotic relationship with the roots of a plant. They draw their nutrients from the plant whilst at the same time releasing nutrients from the soil for the benefit of the plant.

Mycotoxin: microbial metabolite that is toxic to mammals, particularly man.

Nematodes: microscopic worms some of which parasitize insects and plants, whilst others are free-living feeding on bacteria and fungi.

Node: the point on the stem of a plant at which one or more shoots, leaves or flowers are attached.

Nucleotide: the structural unit of a nucleic acid comprising a nucleoside (base and sugar) linked to a phosphate.

Organelle: a cellular structure which has specialized functions within the cells such as the chloroplast or mitochondria.

Oviposition: the act of depositing eggs.

Parasitoid: an organism which spends most of its life cycle within a host and ultimately kills the host. A parasite may live within the host without eventually killing it.

Pathogen: a parasite such as a microorganism or an insect which causes a disease within its host.

Pericarp: the wall of a ripe ovary or fruit.

Phenotype: the observable characteristics which define a particular organism.

Phenylpropanoid: a natural product which is biosynthesized by the shikimate pathway and which contains a phenyl group attached to a three-carbon chain as exemplified by cinnamic acid.

Pheromone: a substance or a mixture which is produced by a living organism that conveys a message to other members of the same species. For example there are aggregation pheromones, alarm pheromones, sex pheromones, territorial pheromones and trail pheromones.

Phloem: the cells and fibres which transport organic substances, principally sugars, from the leaves to the sites of storage such as the roots. The xylem carries water and nutrients in the opposite direction from the roots.

Photoperiodism: the response of plants to rhythmic changes in the relative length of night and day and of light intensity. A 'long-day' plant is one that requires a long photoperiod with less than twelve hours of darkness in twenty fours hours to induce flowering (summer flowering plants), whilst a 'short-day' plant forms flowers when the night is longer than the day (spring and autumn flowering plants).

Photosynthesis: the physical process by which plants capture the energy of the sunlight and use this in a chemical reaction to convert carbon dioxide and water to sugars.

Phytoalexin: a compound produced by a plant in response to microbial attack. These are distinct from phytoanticipins which are pre-existing antimicrobial compounds found in plants.

Phytochrome: a protein–tetrapyrrole complex which absorbs red light and mediates various plant circadian and other responses.

Phytotoxin: a microbial metabolite that is toxic to plants.

Plant growth regulators: a generic term for synthetic and natural products which regulate the growth of plants in a hormonal manner.

Plastids: major organelles found in plants which are responsible for photosynthesis (chloroplasts) and the synthesis of other compounds such as terpenes (leucoplasts), pigments (chromoplasts) and the storage of starch (amyloplasts).

Poaceae: a family of plants formerly known as the Gramineae, which are grasses and include the economically important cereals such as wheat, corn and barley.

Polyketide: a compound which is biosynthesized by the linear poly-merization of acetate units.

Proboscis: the projecting parts of the mouth of an insect which are used for sucking in food.

Protoplast: the contents of the living cell within the cellular membrane or plasmalemma.

Rhizome: a specialized underground stem of a plant which can serve as a storage organ and as a means of spreading the plant.

Rosaceae: a family of plants which includes not only the roses but also a number of edible fruits in the genus *Prunus* (plums) and *Malus* (apples).

Sapwood: the outer wood of a tree trunk comprising the xylem tissues and fibres involved in water transport.

Saprophyte: an organism such as a fungus which decomposes dead material externally and then absorbs the resultant nutrients.

Semiochemical: a generic term for chemicals that are involved in car-rying a message. It is normally used in the context of insect chemistry to include pheromones, allomones and kairomones.

Senescence: the process of deterioration preceding the death of a plant organ such as a leaf.

Solanaceae: a family of plants which includes the potato, the tomato and peppers, as well as deadly nightshade (*Belladonna*), tobacco (*Nicotiana*) and petunia (*Petunia* sp.)

Species: a basic taxonomic unit of genetically related, morphologically similar interfertile individuals. There can be sub-species and cultivars.

Spore: a small propagule of a bacterium or fungus having the function of a seed.

Stylet: the primitive mouth of some nematodes and aphids which is adapted for piercing cell walls.

Stamen: the male reproductive organ of a flower.

Symbiosis: the relationship of two organisms living together in either a mutually beneficial or parasitic relationship.

Synomone: a substance which is released by a member of one species which has a mutually beneficial effect on a member of another species.

Tachinid fly: flies of the Tachinidae family whose larvae are internal parasites of other insects.

Tap root: the central root which forms the main axis of the root system.

Teratogen: a compound which, when ingested by a pregnant mammal, leads to deformed off-spring.

Terpenoid: a natural product derived by the head-to-tail polymerization of C_5 isoprene (isopentenyl diphosphate) units. A monoterpenoid comprises two isoprene units, a sesquiterpenoid contains three units, a diterpenoid has four units, a sesterterpenoid has five units, a triterpenoid has six units and a carotenoid has eight units.

Tuber: swollen portions of the plant, usually the roots system, which act as storage organs.

Vascular system: the combined strands of the xylem and phloem which are responsible for the transport of water, nutrients and the products of metabolism within the plant.

Vacuole: a membrane-bound body within the cytoplasm of the cell.

Vector: an insect or animal which transmits a disease causing organism from one plant to another.

Xylem: the part of the vascular system which is responsible for the transport of water and dissolved nutrients from the roots.

Yeast: a fungus-like eukaryotic microorganism such as *Saccharomyces cerevisiae* which are often found on fruit and which are can be responsible for the fermentation of sugars to form ethanol.

Subject Index

CPSIA information can be obtained at www.ICGtesting.com
Printed in the USA
LVOW071435270912

300606LV00008B/14/P